地学英语阅读立体化系列教材
中央高校教育教学改革基金(本科教学工程)资助

地学精要英语阅读教程

English Reading for Essentials of Geosciences

杨红燕　陈　俐　赖小春　主　编
汪卫红　胡冬梅　王晓婧　周诗羽　任利民　副主编

图书在版编目(CIP)数据

地学精要英语阅读教程/杨红燕,陈俐,赖小春主编.—武汉:中国地质大学出版社,2022.12
地学英语阅读立体化系列教材/赖旭龙,董元兴,刘芳,赖小春主编.
ISBN 978-7-5625-5428-8

Ⅰ.①地⋯　Ⅱ.①杨⋯　②陈⋯　③赖⋯　Ⅲ.①地球科学-英语-阅读教学-高等学校-教材　Ⅳ.①P

中国版本图书馆CIP数据核字(2022)第198354号

地学精要英语阅读教程			杨红燕　陈　俐　赖小春	**主　编**
			汪卫红　胡冬梅　王晓婧　周诗羽　任利民	**副主编**

责任编辑:龙昭月	选题策划:张　琰　龙昭月	责任校对:何澍语
出版发行:中国地质大学出版社(武汉市洪山区鲁磨路388号)		邮政编码:430074
电　　话:(027)67883511	传　　真:(027)67883580	E-mail:cbb@cug.edu.cn
经　　销:全国新华书店		http://cugp.cug.edu.cn
开本:850毫米×1168毫米 1/16		字数:324千字　印张:11.5
版次:2022年12月第1版		印次:2022年12月第1次印刷
印刷:武汉市籍缘印刷厂		
ISBN 978-7-5625-5428-8		定价:45.00元

如有印装质量问题请与印刷厂联系调换

"地学英语阅读立体化系列教材"

编撰委员会

主　编　赖旭龙　董元兴　刘　芳　赖小春
顾　问　唐辉明　刘勇胜　周爱国　马昌前　殷坤龙
　　　　胡祥云　李建威　马　腾　章军锋　Roger Mason
　　　　Timothy Kusky　王　璐
编　委　（按姓氏笔画排序）
　　　　王　伟　王国念　王晓婧　卢　云　付　蕾　冯　迪
　　　　任利民　刘　敏　刘春华　刘晓琴　刘倩倩　刘雪莲
　　　　江　翠　严　瑾　杨红燕　肖　珊　何　霜　汪卫红
　　　　张　莉　张红燕　张伶俐　张基得　陈　俐　陈晓斌
　　　　周宏图　周诗羽　赵　妍　胡冬梅　胡志红　姚夏晶
　　　　秦　屹　黄曼丽　彭　林　葛亚非　曾艳霞　蓝希君

《地学精要英语阅读教程》

主　编　杨红燕　陈　俐　赖小春
副主编　汪卫红　胡冬梅　王晓婧　周诗羽　任利民

Preface

I welcome CUG Press's publication of a volume of Earth Science texts in English accompanied by readings by native speakers, and I urge you to listen, read texts aloud, record them and listen and compare yourself.

English is the international language of Earth Science used to publish results in international journals and exchange ideas at meetings all round the world. Nowadays Earth Science graduates from Chinese universities including CUG have English language skill in reading, but are behind non-native speakers from Europe, the Americas and the Indian sub-continent in listening, speaking and writing. I have seen an immense improvement at CUG over the last 30 years, but my experience in classes, seminars, meetings, thesis defences and editing the English of papers submitted to CUG's *Journal of Earth Sciences* convinces me that further improvement is needed.

Search for the meaning when you read silently. Begin by reading quickly ("skimming") to get the general meaning of a paragraph and write down in English what you judge to be the main meaning. Then read again more carefully. The structure of English sentences is different from Chinese because a writer will begin with the main subject and give evidence in the sentence complement afterwards. For example, a native speaker might write "a P-T evolution diagram of the granites has been drawn using calculated crystallization temperatures and temperature changes during magma evolution", rather than "based on the calculated temperatures of the granites and temperature changes during magma evolution, a P-T evolution diagram of granite can be drawn".

Spoken English uses voice tones differently from Chinese. Listen to the recordings of the texts for examples. You should already know that a rising tone at the end of a sentence signals a question, "Is this the way to the rock mechanics *láb*?" Voice tone is often raised for emphasis and will help you to recognise key words and phrases as you listen to a talk. "There was a *major change* in Earth's tectonics about 2.1 billion years ago." Falling tones indicate disagreement, "Professor, I think your theory is *wròng*."

Written English derives from speech and you should use every opportunity you can to listen and speak. Local spoken pronunciation varies but at scientific meetings the English is almost always correct, so if you don't understand a native speaker you need to improve your listening to include several varieties. (Be tolerant of non-Chinese non-native English speakers who face the same difficulties as you.) Chinese learners of English worry about differences between American and British English but they are small compared with the varieties of English spoken in different parts of the United States and the British Isles, and of course there are also distinctive types of English in Canada, Australia, New Zealand, and South Africa. Millions of people in India, Pakistan, Sri Lanka, Bangladesh, Singapore, and some African countries speak English as their first language but have never used it to converse with a British or an American person! We have included examples of non-standard English pronunciation among our readings.

Relax while you are learning and enjoy the texts and readings while you learn from them. Don't worry about making mistakes because your teachers will help you to put them right. The exercises in the text let you practise scientific and language points. Your improved fluency in English will raise your understanding of Earth Science and open up a whole new world of the English communication.

Roger Mason

2 February 2018

Roger Mason was born in London, England on May 4, 1941. He received a BA degree from Cambridge University in 1963, a PhD in 1967 and a Certificate of Education from the University of Hertfordshire, England in 1990. From 1966 to 1989 he taught Metamorphic Petrology, Tectonics and Field Geology at University College London; Birkbeck College, London University; and the University of Bedfordshire, England. He taught a special course in Metamorphic Petrology at Wuhan College of Geology in 1986 and bilingual classes in Metamorphic Geology at China University of Geosciences (Wuhan) (CUG Wuhan) from 1996 to 2011. He is a professor of Mineralogy and Petrology at CUG (Wuhan). His textbook *Petrology of the Metamorphic Rocks* was published in the United Kingdom (first edition in 1978, second edition in 1990), and his textbook *Metamorphic Geology* with co-author Professor Sang Longkang by China University of Geosciences Press (2007). His publications are regularly cited both in Britain and China. He won "the China International Friendship Prize" in 2001, the highest prize awarded by the Chinese government to foreign experts, for his contribution to bilingual teaching and the Sino-British international cooperation. He lives in retirement in London, England and continues to collaborate with Chinese colleagues in research.

序

随着中国国际地位的攀升、"一带一路"倡议的推进,国际交流与合作的机会越来越多,中国制造和中国标准亦初见端倪。中国要走向国际,需要越来越多的技术输出,对既懂专业又懂外语的专业技术人才的需求也更为迫切。传统上以通用语言能力为主的大学英语教学活动难以满足国际化专业技术人才的语言需求。因此,不少大学开始改革大学英语教学,增加适应中国国际化战略的专门用途英语(English for specific purposes, ESP)教学,针对各专业学科编写的ESP教材也相继出版面世。为配合我校地学类双一流学科人才培养建设,我们依托我校丰厚的地学英语专业文献,借助地学专家的学术指导,编撰了这套"地学英语阅读立体化系列教材"。

该套教材分为4册,遵循由简入深的方式,按照ESP教学类别,对所选材料进行分门别类,便于学习者逐级学习使用。教材前两册,即《地学英语阅读教程》和《地学精要英语阅读教程》,以一般性学术英语(English for general academic purpose, EGAP)为原则编撰。作为学术文献阅读的基础,重点介绍地学基础类文章,使学习者对地学类英语文献逐步形成系统的认知,加深对地学类英语文章的文本特色理解,并积累一定的学术文献阅读技巧,为后续两册的学习打下坚实的语言基础,为更高级专业地学学术文献阅读作铺垫。该套教材后两册,即《地学文献阅读》和《地学语篇语用》,供后续学术英语阅读教学使用,以专用学术英语(English for specific academic purpose, ESAP)为原则编撰。教材摘取真实的学术期刊论文或报告作为阅读文本,使学习者学会快速阅读本专业文献,逐步了解专业文献的写作特点和写作技巧,辅助学生进行专业学术论文写作。通过本系列教材的学习,学习者可以完成由EGAP阅读到ESAP阅读的平稳过渡,既能提高阅读专业文献的能力,又能为英语论文写作打下一定的基础。

本系列教材适合作为大学英语的学术英语阅读课程,每册可满足每学期24~32学时的教学需要,也可以作为地学专业双语教学的阅读教材。同时,各册教材重点突出,定位明确,可以作为英语选修课教材,供不同需求层次的学生单独使用。

《地学英语阅读教程》——**Basic,基础篇**。EGAP阅读基础版,本册以地学基础类文章为阅读文本,浅显易懂,以阅读了解地学类基础知识为目标。

《地学精要英语阅读教程》——**Progressive,拓展篇**。EGAP阅读文本,重点选取与地学相关的专业文本,文体简洁,知识易懂。学习者通过学习本册教程,可以对地学涵盖的相关领域有一个系统的认知。

《地学文献阅读》——**Advanced,提高篇**。ESAP阅读文本,重点选取了地学专业学术文章或学术报告,用词专业性强,强调专业文本的精确性、简洁性,以及文本结构的严谨性。学习者通过本册教程的学习,可以对地学专业类的学术文献有更直接的感性认识,加深对学术文本语言特色和写作规范的理解。

《地学语篇语用》——**Proficient,专业篇**。ESAP加强型阅读文本,选取地学专业期刊论文或学术报告,用词专业性强。本册教程强调学术文献语篇和语言使用的理解。通过语篇建构和语言使用剖析，以及阅读和写作技巧的训练，学习者可以依据论文语篇特色，尝试撰写英文学术研究论文和英文学术研究报告。

本系列教程可以4册连用，循序递进；也可以根据不同层次学生的需求选取一两册满足教学需要，其他教材作为辅助拓展阅读材料，供学生自学。

由于地学英语学术文本的特殊性，专业词汇难以上口，我们在每册书的后面都提供了加注音标的词汇总表，便于学习者查阅朗读。本系列教材还提供立体化教材资源服务，既有纸质版教材书本便于携带阅读，也有MOOC视频作为教学辅导，每个章节处的二维码链接有语音资源，勒口处还配有教学参考PPT资源，扫码可见。

此系列教材只是我们进行大学英语ESP教学改革的第一步，还会面临各种挑战。我们相信，只要我们加倍努力，不断提高学术英语和地学专业知识素养，不断更新改进我们的学术英语阅读教材，一定会为ESP的教学打开一片天地。

2017年12月

前　言

专门用途英语(ESP)教学已成为我国高校,特别是理工类高校大学英语教学改革及发展的方向。《地学精要英语阅读教程》作为地学英语阅读系列教材的拓展篇,延续了《地学英语阅读教程》的 ESP 教学理念,旨在帮助学习者对地球科学基本知识形成系统认知,加深对地学英语文本特色的熟悉度,并积累一定的地学专业文献阅读技巧。

本教材具有以下特色：

(1)注重地学专业体系与语言能力的有机融合。本教材以"大地学"的专业知识体系为编写主线,同时兼顾语言知识,实现语言学习与专业内容学习的有机融合;在编写体例上,既涵盖专业词汇和术语,又兼顾英语句法结构分析,部分练习直接对接托福考试的阅读理解。

(2)促进通用语言能力向专业英语能力的过渡。本教材以阅读技能训练为主,着重增强"大地学"专业学生的学术英语阅读能力和国际交流能力,实现从通用英语学习向专业英语学习的过渡,满足国际化专业技术人才的语言学习需求。

(3)为教材使用者提供立体化的教学方案。为了最大限度地满足教师课堂教学和学生课下学习的需要,各单元均可以通过扫描二维码的方式获取词汇和课文的音频、课后练习答案、教学课件及相关电子资源。

《地学精要英语阅读教程》共 8 个单元,每个单元由 A 篇、B 篇、C 篇 3 篇课文以及围绕课文设计的练习组成,各单元的结构如下。

(1)单元概览(Preview)：以 80～100 词的语篇引导读者进入单元话题,并对本单元 3 篇阅读文章的主题进行提炼,方便读者综合了解单元内容。

(2)A 篇课文(Passage A)：节选改编自英语原版教材、专业期刊或专业网站的文章,长度控制在 1200～1500 词之间,重在展现各单元主题领域的基础知识,文后配有相应的阅读理解练习。

(3)B 篇课文(Passage B)：选自历年托福考试真题或者模拟题,是结合托福题型对相关主题的阅读拓展,长度控制为 600～800 词。

(4)C 篇课文(Passage C)：与 A 篇课文相似,节选改编自英语原版教材、专业期刊或专业网站的文章,重在呈现该领域的前沿或发展趋势,长度控制在 1200～1500 词之间,文后配有相应的阅读理解练习。

(5)综合练习(Exercises)：由词汇练习(Vocabulary)、句法结构(Sentence Structure)、翻译(Translation)3 个部分练习组成。词汇练习既包含对单元主题相关专业词汇的总结,也包括对通用学术词汇的归纳;句法结构练习除第一单元外,其余单元聚焦于本单元出现的某一特定句型结构,助力读者对英语长难句的理解;翻译练习由英译汉和汉译英两部分组成,英译汉重点训练学生对本单元聚焦句型的理解,汉译英练习着重考察学生对本单元专业词汇的综合运用。

本书是地质、资源、环境、工程等"大地学"相关专业学生从通用大学英语过渡到专业学术英语学习的"专门用途英语"教材,建议教学周期为 1 个学期,学时建议为 24～32 个学时。本教材可以作为"大地学"相关专业本科生和研究生学术英语阅读进阶的主要教材,帮助学生从通用英语学习过渡到专业英语学习,为专业学习尤其是专业学术论文写作打下基础。除适合"大地学"相关专业的本科生和研究生外,本教材还可以作为"大地学"相关专业从业人员的英语自学材料,提高英语专业学生跨学科语言技能的教材,以及有出国需求的各类人士托福备考教材。

本教材的编写团队来自中国地质大学(武汉)外国语学院和地球科学学院:由任利民完成教材各单元的主题框架;由杨红燕、陈俐、赖小春、汪卫红完成单元编写体例设计;杨红燕、赖小春、汪卫红共同承担了第一单元样章的编写;第二至第八单元由杨红燕、陈俐、胡冬梅、赖小春、汪卫红、周诗羽、王晓婧、任利民完成编写;杨红燕、陈俐、赖小春承担了全书的统稿工作;陈俐、任利民承担了教材译文答案的审校工作。

本教材的顺利出版得到了中国地质大学(武汉)外国语学院和本科生院的大力支持,本教材顾问团队和编委为本教材的编写提供了宝贵建议,汪卫红、王国念、唐慧君、冯迪为本教材提供了单词和课文的录音,在此一并表示衷心的感谢。

由于编写时间仓促,编者水平有限,本教材难免会有不尽如人意之处,我们真诚希望使用本教材的广大教师、学生和其他专业人士给我们指出,以便我们更正和改进。

编 者
2022 年 9 月

Contents

Unit 1 Geoscience *1*

    Passage A The Emergence and Evolution of the Earth System Science *1*

    Passage B Seismic Waves and the Earth's Core *8*

    Passage C Convection Plumes in the Lower Mantle *13*

    Exercises *18*

Unit 2 Geomaterials *23*

    Passage A Rock Forming Minerals *23*

    Passage B Physical Properties of Mineral *28*

    Passage C Attributes of Geomaterials in Construction *33*

    Exercises *38*

Unit 3 Atmospheric and Climate Sciences *41*

    Passage A Global Climate Change *41*

    Passage B Climate and Urban Development *46*

    Passage C ENSO Events and Climate Change *51*

    Exercises *56*

Unit 4 Soil Science *60*

    Passage A A Brief Introduction to Soil *60*

    Passage B Minerals and Plants *66*

    Passage C What is Soil Health *71*

    Exercises *75*

Unit 5 Modern Hydrology and Marine Sciences *78*

    Passage A The Hydrologic Cycle *78*

Passage B   Glacier Formation   *82*
Passage C   Ocean Water   *87*
Exercises   *90*

Unit 6   Resources and Energy on the Earth   *94*

Passage A   Natural Resources: Classification and Protection   *94*
Passage B   The Earth's Energy Cycle   *99*
Passage C   Nuclear Energy   *104*
Exercises   *109*

Unit 7   Low Carbon Economics   *113*

Passage A   Energy, Society and Environment   *113*
Passage B   Smart Energy   *119*
Passage C   Geoscience and Decarbonization: Current Status and Future Directions   *123*
Exercises   *128*

Unit 8   Natural Disasters   *132*

Passage A   Estimating and Controlling Floods   *132*
Passage B   Volcano Monitoring   *138*
Passage C   Mitigation of the Impacts on the Earth by Near-Earth Objects   *142*
Exercises   *147*

Glossary   *151*

Unit 1 Geoscience

Geoscience, also called Earth Science, is the study of the Earth. It is a highly interdisciplinary field, combining elements of geology, oceanography, geography, and so on. In this unit, you will read about: 1) the Earth system science (ESS), a rapidly emerging transdisciplinary endeavor aiming at understanding the structure and functioning of the Earth as a complex, adaptive system, 2) the use of earthquake waves to probe the Earth's core, and 3) possible explanation for the volcanoes in the middle of a plate.

Passage A

The Emergence and Evolution of the Earth System Science*

Beginnings (pre-1970s)

❶ Past conceptualizations of the Earth formed important **precursors** to the contemporary understanding of the Earth system. Examples include J. Hutton's 1788 *Theory of the Earth*, Humboldtian science in the 19th century and Vernadsky's 1926 *The Biosphere*. Understanding the historical roots of the Earth system science (ESS), however, requires a focus on the second half of the 20th century when, in a Cold War context, important shifts occurred in the Earth and environmental sciences.

precursor *n.* 先驱,前导
epitomize *v.* 成为……的典范,作为……的缩影

❷ In the middle of the 20th century, international science started to develop, **epitomized** by the International Geophysical

---

* The passage is adapted from STEFFEN W, RICHARDSON K, ROCKSTRÖM J, et al., 2020. The emergence and evolution of the Earth system science[J]. Nature Reviews the Earth & Environment, 1: 54–63.

geosphere *n.* 岩石圈,陆界
glaciology *n.* 冰河学,冰川学
oceanography *n.* 海洋学
meteorology *n.* 气象状态,气象学
climatological *adj.* 与气候学有关的
instrumentation *n.* 使用仪器,仪表化
numerical *adj.* 数字的,用数字表示的
paradigm *n.* 典范,范例,样板,范式
ecology *n.* 生态学
biosphere *n.* 生物圈
ozone *n.* 臭氧,臭氧层(ozone layer 的简写)
depletion *n.* 损耗,耗尽
finitude *n.* 有限,界限,限制
entity *n.* 实体,存在
ensemble *n.* 共同,一起,集合,整体
hypothesize *v.* 假设,假定
homeostatic *adj.* 自我平衡的,原状稳定的

Year (IGY)[1] 1957-1958. This unprecedented research campaign coordinated the efforts of 67 countries to obtain a more integrated understanding of the **geosphere**, particularly **glaciology**, **oceanography**, and **meteorology**. One of the key impacts of the IGY was a lasting transformation in the practices used to understand how the Earth works. The interpretative and qualitative geological and **climatological** research based on field observations—as classically studied by geographers—was replaced by field **instrumentation**, continuous and quantitative monitoring of multiple variables and **numerical** models. This transformation led to the two contemporary **paradigms** that structure Earth Science: modern climatology and plate tectonics.

❸ **Ecology** and environmental sciences also developed rapidly. Ecosystem ecology emerged with the work of G. E. Hutchinson[2] and the brothers H. Odum and E. Odum[3], supported by the Scientific Committee on Problems of the Environment (SCOPE). Large projects such as the International Biological Program (IBP)[4] were a major step towards a global ecological study. These efforts provided the basis for understanding the role of the **biosphere** in the functioning of the Earth system as a whole.

❹ The 1960s and 1970s were marked by a broadening cultural awareness of environmental issues in both the scientific community and the general public. Driving this increased awareness were the publication of R. Carson's *Silent Spring*, "the Only One Earth" discourse at the 1972 United Nations Conference on the Human Environment, the first alerts on **ozone depletion** and climatic change and the Club of Rome's 1972 publication of the *Limits to Growth* (report), the latter warning of the **finitude** of economic growth due to resource depletion and pollution. Visual images of the Earth, in particular *The Blue Marble* image taken by the crew of the Apollo 17 spacecraft on 7 December 1972, sharpened the research focus on the planet as a whole and highlighted its vulnerability to the general public.

❺ Amidst these developments, J. Lovelock introduced the term "Gaia"[5] in 1972 as an **entity** comprised of the total **ensemble** of living beings and the environment with which they interact and **hypothesized** that living beings regulate the global environment by generating **homeostatic** feedbacks. Although this hypothesis generated scientific debate and criticism, it also generated a new way of thinking about the Earth: the major influence of the **biota** on the global environment, the importance of the **interconnectedness** and feedbacks that link major components of the Earth system.

❻ The scientific developments up to 1980—from Vernadsky's pioneering research, through large-scale field campaigns and the emerging environmental awareness of the 1970s, to Lovelock's Gaia—led to a new understanding of the Earth, challenging a purely **geophysical** conception of the planet and transforming our view of the environment and nature.

Founding a new science (1980s)

❼ Triggered by the growing recognition of global changes such as human-driven ozone depletion and climatic change, a series of workshop and conference reports in the 1980s called for a new "science of the Earth". The calls were based on the acknowledgement that if a new science was to be founded, it would need to be based on the newly emerging recognition of the Earth as an integrated entity: the Earth system.

❽ At National Aeronautics and Space Administration (NASA), the new scientific endeavor was named "the Earth system Science". The NASA Earth system science Committee was established in 1983 and aimed at supporting the Earth Observing System (EOS) satellites and associated research that helped drive the evolving definition of ESS via observations, modelling and process studies. The NASA-led research **initiatives** also developed new visual representations of the Earth system, most famously the NASA Bretherton Committee diagram (1986). The Bretherton diagram (Figure 1) was the first systems-**dynamics** representation of the Earth system to couple the physical climate system and **biogeochemical** cycles through a complicated array of forcings and feedbacks. The Bretherton diagram epitomized the rapidly growing field of ESS through its visualization of the interacting physical, chemical, and biological processes that connect components of the Earth system and through the recognition that human activities were a significant driving force for change in the system. This created a significant challenge in bringing different disciplines together to study the Earth system as a whole.

biota *n.* 生物区(系)，一时代(一地区)的动植物
interconnectedness *n.* 互联性
geophysical *adj.* 地球物理学的
initiative *n.* 措施，倡议，主动性，积极性
dynamics *n.* 动力学，力学，动力
biogeochemical *adj.* 生物地球化学的

terrestrial *adj.* 地球的,地球上的
hydrological *adj.* 水文学的
locus *n.* 地点,所在地
disciplinary *adj.* 学科的
convergence *n.* 趋同,汇集,相交
transdisciplinary *adj.* 跨学科的,学科间的

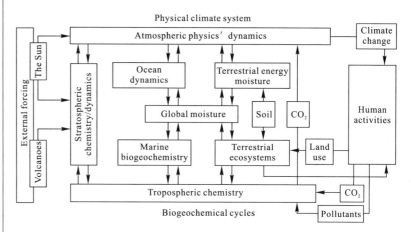

Figure 1　The NASA Bretherton diagram of the Earth system

❾　The challenge of international commitment and disciplinary integration was addressed in 1986 by the International Council for Science (ICSU) with the formation of the International Geosphere-Biosphere Program (IGBP), which joined the World Climate Research Program (WCRP), formed in 1980 to study the physical-climate component of the Earth system. The IGBP was originally structured around a number of core projects on biogeochemical aspects of the Earth system: ocean carbon cycle, **terrestrial** ecosystems, atmospheric chemistry, the **hydrological** cycle, and others. Two projects of particular importance were Past Global Changes (PAGES) and Global Analysis, Integration, and Modelling (GAIM), given their **locus** of strong **disciplinary** integration. In addition, the IGBP developed a dedicated project on data and information systems (DIS), especially remotely sensed data, to support the research.

❿　This **convergence** of disciplines accelerated the evolution of ESS, evident as a transition from isolated process studies to interactions between these processes and increasingly global-level observations, analyses and modelling. ESS thus facilitated the transformation from interdisciplinary research (where multiple disciplines work together to tackle common problems) to **transdisciplinary** research (where disciplinary boundaries fade as researchers work together to address a common problem).

Going global (1990s – 2000s)

⑪ The formal launch of the IGBP in 1990 and the widespread use of the Bretherton diagram powered the ongoing development of ESS. Thus, motivated by a suite of studies that illustrated the importance and relevance of ecological research to climate change, **biodiversity** and **sustainability** more broadly, the international research program DIVERSITAS was created in 1991 to study the loss of, and change in, global biodiversity, complementing the IGBP's research on the functional aspects of terrestrial and marine ecosystems.

⑫ In 1996, the International Human Dimensions Program (IHDP) on Global Environmental Change was founded, providing a global platform for social science research that explored the human drivers of change to the Earth system and the consequences to human and societal wellbeing. This global system of international research programs, including the WCRP, IGBP, DIVERSITAS, and IHDP, provided "workspaces" for international scientists of different disciplines to come together, which was critical for the development of ESS. In the early 2000s, this more complete suite of global-change programs, along with the emerging concept of sustainability, would give birth to sustainability science.

⑬ In the late 1990s, H. J. Schellnhuber introduced and developed two concepts that were fundamental for ESS in his paper "Earth system analysis and the second Copernican revolution": the dynamic, **co-evolutionary** relationship between nature and human civilization at the planetary scale and the possibility of **catastrophe** domains in the co-evolutionary space of the Earth system. The first provided the conceptual framework for fully integrating human dynamics into an Earth-system framework. The second introduced the risk that global change may not unfold as a linear change in the Earth-system functioning but, rather, that human pressures could trigger rapid, **irreversible** shifts of the system into states that would be catastrophic for human well-being. Indeed, the discovery of the **stratospheric** ozone hole showed that humanity, by luck rather than design, has already narrowly escaped the creation of a catastrophe domain.

⑭ Over a critical 5-year period from 1999 through 2003, the

biodiversity *n.* 生物多样性
sustainability *n.* 持续性，能维持性
co-evolutionary *adj.* 协同进化的
catastrophe *n.* 灾难，灾祸，困境
irreversible *adj.* 不可挽回的，无法逆转的
stratospheric *adj.* 平流层的，同温层的

underpin v. 支持,巩固,构成……的基础

Anthropocene n. 人类世

IGBP accelerated its transition from a collection of individual projects to a more integrated ESS program, with the 1999 IGBP Congress being the key to achieving the required integration. The Congress launched both the IGBP synthesis project in 2004 and a major international conference in 2001. The synthesis project resulted in the publication of *Global Change and the Earth System*, an integrator of a vast amount of global-change research. It also provided the scientific basis for the *Amsterdam Declaration*[6] and emphasized research that would **underpin** the new concept of the **Anthropocene** proposed by Crutzen in 2000.

⑮ The 2001 conference, Challenges of a Changing Earth—co-sponsored by the four international global-change programs (the IGBP, WCRP, IHDP, and DIVERSITAS)—was truly international, attracting 1,400 participants from 105 countries. The conference introduced the *Amsterdam Declaration*, triggering the formation of the Earth System Science partnership (ESSP) to connect fundamental ESS with issues of central importance for human well-being: food, water, health, carbon, and energy.

Contemporary ESS (beyond 2015)

⑯ By 2015, ESS was well established and the time was right for a major institutional restructure built on a higher level of integration. Indeed, the IGBP, IHDP, and DIVERSITAS were merged in 2015 into the new program, Future Earth, aiming to accelerate the transformation to global sustainability through research, and innovation. It builds on the research of the earlier global-change programs but works more closely with the governance and private sectors from the outset to co-design and co-produce new knowledge towards a more sustainable future. Meanwhile, the WCRP continued, along with some IGBP core projects, such as the International Global Atmospheric Chemistry (IGAC) project, PAGES and the ESSP Global Carbon Project.

⑰ A broad range of research centers now directed their work towards ESS and global sustainability research; for example, the Potsdam Institute for Climate Impact Research (PIK), the US National Center for Atmospheric Research (NCAR), the Stockholm Resilience Centre (SRC), and the International Institute for Applied Systems Analysis (IIASA). Although universities maintained their traditional discipline-based faculties, as the emphasis on interdisciplinarity and global-level studies grew, interdisciplinary ESS programs also emerged in many universities around the world. The revolution in digital communication links these and many other research bodies in an expanding global ESS effort.

Notes

(1) **The International Geophysical Year (IGY)** was a worldwide program of geophysical research that was conducted from July 1957 to December 1958. It was directed toward a systematic study of the Earth and its planetary environment.

(2) **George Evelyn Hutchinson** (January 30, 1903–May 17, 1991), was a British ecologist sometimes described as the "father of modern ecology". He is known as one of the first to combine ecology with mathematics. He became an international expert on lakes and wrote the four-volume *Treatise on Limnology* in 1957.

(3) **Howard T. Odum (1924–2002) and Eugene P. Odum (1913–2002)** were leading figures in the development of ecosystem ecology. Their father, the sociologist Howard Washington Odum, was a leading organicist thinker. From their father, Eugene and Howard took the idea of the integration of parts to form a larger social whole, which they later expanded in their holistic ecosystem thinking.

(4) **The International Biological Program (IBP)** was an effort between 1964 and 1974 to coordinate large-scale ecological and environmental studies. Organized in the wake of the successful International Geophysical Year (IGY) of 1957–1958, the International Biological Program was an attempt to apply the methods of big science to ecosystem ecology and pressing environmental issues.

(5) **The Gaia hypothesis** was formulated by the chemist James Lovelock and co-developed by the microbiologist Lynn Margulis in the 1970s, which proposes that living organisms interact with their inorganic surroundings on the Earth to form a synergistic and self-regulating, complex system that helps to maintain and perpetuate the conditions for life on the planet. Lovelock named the idea after Gaia, the primordial goddess who personified the Earth in Greek mythology.

(6) ***Amsterdam Declaration*** described the key findings of a decade of the Earth system science. The focus was on recognizing the Earth as a single system with its own inherent dynamics and properties at the planetary level, all of which are threatened by human-driven global change.

Reading Comprehension

 Keys

Directions Fill in blanks according to Passage A.

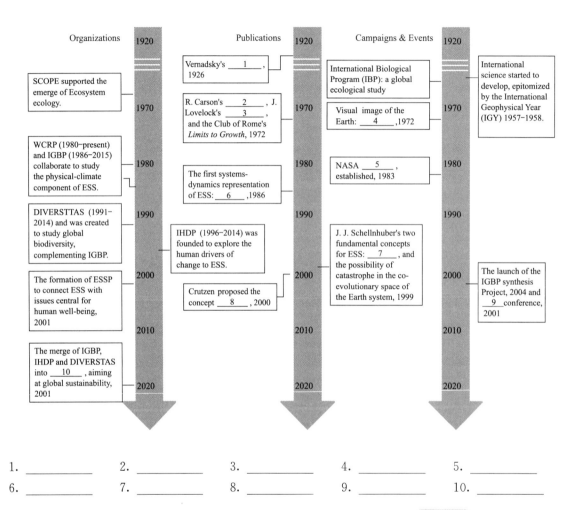

1. _____ 2. _____ 3. _____ 4. _____ 5. _____
6. _____ 7. _____ 8. _____ 9. _____ 10. _____

Passage B

 Scan and read along the vocabulary

seismic *adj.* 地震的，由地震引起的

 Scan and read along the passage

Seismic Waves and the Earth's Core*

❶ Knowledge of the Earth's deep interior is derived from the study of the waves produced by earthquakes, called **seismic** waves.

---

* This passage is adapted from 语言研究中心，2011.托福官方指南考点详解：基础篇（阅读分册）[M].北京：北京语言大学出版社.

Among the various kinds of seismic waves are primary waves (P-waves) and secondary waves (S-waves)[1]. Primary and secondary waves pass deep within the Earth and therefore are the most instructive. Study of abrupt changes in the characteristics of seismic waves at different depths provides the basis for a threefold division of the Earth into a central core; a thick, overlying mantle; and a thin, enveloping crust. Sudden changes in seismic wave **velocities** and angles of transmission are termed discontinuities.

❷ One of the discontinuities is the Gutenberg discontinuity[2], which is located nearly halfway to the center of the Earth at a depth of 2,900 kilometers and marks the outer boundary of the Earth's core. At that depth, the S-waves cannot **propagate**, while at the same time P-wave velocity is **drastically** reduced. S-waves are unable to travel through fluids. Thus, if S-waves were to encounter a fluid-like region of the Earth's interior, they would be absorbed there and would not be able to continue. Geophysicists believe this is what happens to S-waves as they enter the outer core. As a result, the S-waves generated on one side of the Earth fail to appear at **seismograph** stations on the opposite side of the Earth, and this observation is the principal evidence of an outer core that behaves as a fluid. Unlike S-waves, P-waves are able to pass through liquids. They are, however, abruptly slowed and sharply refracted (bent) as they enter a fluid medium. Therefore, as P-waves encounter the molten outer core of the Earth, their velocity is reduced, and they are **refracted** downward.

❸ The **radius** of the core is about 3,500 kilometers. The inner core is solid and has a radius of about 1,220 kilometers, which makes this inner core slightly larger than the Moon. Most geologists believe that the inner core has the same composition as the outer core and that it can only exist as a solid because of the enormous pressure at the center of the Earth. Evidence of the existence of a solid inner core is derived from the study of hundreds of records of seismic waves produced over several years. These studies showed that the inner core behaves seismically as if it were a solid.

❹ The Earth has an overall density of 5.5 grams per **cubic centimeter**, yet the average density of rocks at the surface is less than 3.0 grams per cubic centimeter. This difference indicates that

velocity *n.* 速度
propagate *v.* 传播
drastically *adv.* 彻底地,激烈地
seismograph *n.* 地震仪
refract *v.* 使（光线）折射
radius *n.* 半径
cubic centimeter 立方厘米

nickel *n.* (金属)镍
pressurize *v.* 增压,施加压力
silicon *n.* (化学元素)硅
sulfur *n.* 硫磺
meteorite *n.* 陨石
shatter *v.* (使)破碎,碎裂
plausible *adj.* 合理的
magnetic *adj.* 磁的,磁性的,磁化的
magnetic field 磁场
induce *v.* 诱发,因某种原因导致

materials of high density must exist in the deep interior of the planet to achieve the 5.5 grams per cubic centimeter overall density. Under the extreme pressure conditions that exist in the region of the core, iron mixed with **nickel** would very likely have the required high density. Laboratory experiments, however, suggest that a highly **pressurized** iron-nickel alloy might be too dense and that minor amounts of such elements such as **silicon**, **sulfur**, carbon, or oxygen may also be present to lighten the core material.

❺ Support for the theory that the core is composed of iron (85 percent) with lesser amounts of nickel has come from the study of **meteorites**. Many of these samples of solar system materials are iron meteorites that consist of metallic iron alloyed with a small percentage of nickel. Some geologists suspect that iron meteorites may be fragments from the core of a **shattered** planet. The presence of iron meteorites in our solar system suggests that the existence of an iron-nickel core for the Earth is **plausible**.

❻ There is further evidence that the Earth may have a metallic core. Anyone who understands the functioning of a compass is aware that the Earth has a **magnetic field**. The planet itself behaves as if there was a great bar magnet embedded within it. A magnetic field is developed by the flow of electric charges and requires good electrical conductors. Silicate rocks, such as those in the mantle and crust, do not conduct electricity very well, whereas metals such as iron and nickel are good conductors. Heat-driven movements in the outer core, coupled with movements **induced** by the Earth's spin, are thought to provide the necessary flow of electrons (very small particles that carry a negative charge) around the inner core that produces the magnetic field. Without a metallic core, the Earth's magnetic field would not be possible.

Notes

(1) **Primary waves (P-waves) and secondary waves (S-waves)** are two types of seismic wave that moves through the interior of the Earth. The P-wave is a pulse of energy that travels quickly through the Earth and through liquids, moves in a push-and-pull pattern, and causes less damage due to their smaller size. The S-wave follows more slowly only through solids, moves in an up-and-down pattern, and causes more damage due to their greater size.

(2) **The Gutenberg Discontinuity** is situated inside the Earth at a depth of about 2,900 kilometers below the surface, separating the core and the mantle of the Earth.

Reading Comprehension

 Keys

Directions   Answer the following questions according to Passage B.

1. According to Paragraph 1, which of the following findings provides the basis for the Earth's division into three parts?   (   )
   A. All seismic waves pass deep within the Earth.
   B. There are no significant differences in the behavior of P-waves and S-waves.
   C. Discontinuities occur at different depths within the Earth.
   D. The Earth's crust is too thin to account for the size of the planet.

2. According to Paragraph 2, entering the outer core affects seismic waves in each of the following ways EXCEPT   (   )
   A. S-waves do not continue traveling.
   B. S-waves are absorbed by the outer core.
   C. P-waves are slowed down.
   D. P-waves do not change direction.

3. Paragraph 3 supports which of the following statements about the inner core?   (   )
   A. It is smaller in volume than the Moon.
   B. It consists of materials different from those found in the outer core.
   C. It is subject to more pressure than the outer core.
   D. It is not solid even though it behaves seismically as if it were.

4. According to Paragraph 4, why do scientists believe that the Earth's core contains materials other than iron mixed with nickel?   (   )
   A. In the absence of such materials, the core might be too dense.
   B. Iron and nickel do not mix evenly without the presence of lighter elements.
   C. Laboratory experiments showed that minor amounts of silicon, sulfur, carbon, and oxygen are present in samples.
   D. Iron-nickel alloys always contain minor amounts of elements such as silicon, sulfur, carbon, or oxygen.

5. According to Paragraph 5, why does the presence of iron meteorites in our solar system suggest that an iron-nickel core for the Earth is plausible?   (   )
   A. Iron meteorites indicate that our solar system contains large quantities of iron alloyed with nickel.
   B. Some scientists think that the iron meteorites were originally part of the Earth and broke off long ago.
   C. Some scientists think that iron meteorites may be pieces from the core of a planet that broke apart.

D. Iron meteorites in our solar system were formed at about the same time that the Earth was formed.

6. The word "embedded" in Paragraph 6 is closest in meaning to ( )
   A. active.
   B. revolving.
   C. formed.
   D. enclosed.

7. Which of the sentences below best expresses the essential information in the highlighted sentence in the passage?
   A. A combination of different motions produces the Earth's magnetic field, and the magnetic field is thought to cause a flow of electrons around the planet's inner and outer core.
   B. The flow of electrons around the planet that produces the Earth's magnetic field is the combined effect of motions that are thought to occur in the planet's inner core.
   C. The flow of electrons around the inner core that produces the Earth's magnetic field is thought to be the joint result of motions in the outer core and motions caused by the Earth's spin.
   D. Heat-driven motions in the outer core interact with motions caused by the spin of the inner core to produce a flow of electrons that is thought to result in the Earth's magnetic field.

8. In Paragraph 6, why does the author discuss the Earth's magnetic field?
   A. To cast doubt on its existence, given what scientists now know about the Earth's composition.
   B. To provide additional support for the theory of an iron-nickel core for the Earth.
   C. To present a puzzling feature of the Earth that has not yet been explained satisfactorily.
   D. To explain why so many iron meteorites are found near the Earth.

9. The letters [A], [B], [C], and [D] indicate where the following sentence (in bold) could be added to the following part of the passage. Where would the sentence best fit?

   > **Several of these have been named after their discoverers.**

   Knowledge of the Earth's deep interior is derived from the study of the waves produced by earthquakes, called seismic waves. Among the various kinds of seismic waves are primary waves (P-waves) and secondary waves (S-waves). [A] Primary and secondary waves pass deep within the Earth and therefore are the most instructive. [B] Study of abrupt changes in the characteristics of seismic waves at different depths provides the basis for a threefold division of the Earth into a central core, a thick, overlying mantle, and a thin, enveloping crust. [C] Sudden changes in seismic wave velocities and angles of transmission are termed discontinuities. [D]

10. An introductory sentence (in bold) for a brief summary of the passage is provided below. Complete the summary by selecting the THREE answer choices that express the most

important ideas in the passage.

> Our knowledge of the Earth's interior is based on the study of the transmission of seismic waves within the planet.

A. Study of marked changes in the transmission of seismic waves at different depths has led scientists to divide the Earth into the crust, the mantle, and the core.

B. Changes in the characteristics of seismic waves at the Gutenberg discontinuity support the conclusion that the outer part of the Earth's core is liquid.

C. Silicate rocks, which are found in the Earth's mantle and crust, are not good conductors of electricity.

D. Studies of the transmission of seismic waves show that the Earth's core is less dense than the solid crust, which explains why the core exists mostly as liquid.

E. Scientists believe that most of the iron and nickel found in the Earth's core came from iron meteorites scattered into space when the interior of an ancient planet broke apart.

F. A variety of evidence leads to the conclusion that the Earth's inner core is solid and composed primarily of metals, which fits well with the fact the Earth has a magnetic field.

Passage C

Scan and read along the passage

Convection Plumes in the Lower Mantle*

Scan and read along the vocabulary

❶ The concept of crustal plate motion over mantle hotspots has been advanced to explain the origin of the Hawaiian and other island chains and the origin of the Walvis, Iceland-Faroe[1] and other **aseismic ridges**. More recently the pattern of the aseismic ridges has been used in formulating continental reconstructions. I have shown that the Hawaiian-Emperor, Tuamotu-Line and Austral-Gilbert-Marshall Island chains[2] can be generated by the motion of a **rigid** Pacific plate rotating over three fixed hotspots. The motion **deduced** for the Pacific plate agrees with the **palaeomagnetic** studies of seamounts. It has also been found that the relative plate motions deduced from **fault strikes** and **spreading rates** agree with the concept of rigid plates moving over fixed hotspots.

convection n.（热通过气体或液体的）运流，对流
plume n. 羽流，上升之物（mantle plume 地幔柱）
aseismic ridge 无震海岭
rigid (plate) adj. 刚性的（板块）
deduce v. 推论，推断
palaeomagnetic adj. 古地磁的
fault strike 断层走向
spreading rate 扩张速率

---

* This passage is adapted from MORGAN W J, 1971. Convection plumes in the lower mantle[J]. Nature, 230: 42-43.

❷ I now propose that these hotspots are **manifestations** of convection in the lower mantle which provides the motive force for continental drift. In my model there are about twenty deep mantle plumes bringing heat and relatively **primordial** material up to the **asthenosphere** and horizontal currents in the asthenosphere flow **radially** away from each of these plumes. The points of upwelling will have unique **petrological** and **kinematic** properties but I assume that there are no corresponding unique points of downwelling, the return flow being uniformly distributed throughout the mantle. Elsasser has argued privately that highly unstable fluids would yield a thunderhead pattern of flow rather than the roll or convection cell pattern calculated from linear **viscous equations**. The currents in the asthenosphere spreading radially away from each upwelling will produce stresses on the bottoms of the **lithospheric** plates which, together with the stresses generated by the plate-to-plate interactions at **rises, faults**, and **trenches**, will determine the direction in which each plate moves.

❸ Evidently the interactions between plates are important in determining the net force on a plate, for the existing rises, faults, and trenches have a **self-perpetuating** tendency. The plates are apparently quite tough and resistant to major changes because rise crests do not commonly die out and jump to new locations and points of deep upwelling do not always **coincide with ridge crests**. For example, the Galapagos and Reunion upwellings are near **triple junctions** in the Pacific and Indian Oceans. Asthenosphere motion radially away from these hotspots would help to drive the plates from the triple junctions but there is considerable **displacement** between the "pipes to the deep mantle" and the lines of weakness in the lithosphere would enable the plates to move apart. Also, a large isolated hotspot such as Hawaii can exist without splitting a plate in two. I believe it is possible to construct a simple dynamic model of plate motion by making assumptions about the **magnitude** of the flow away from each hotspot and assumptions about the stress/strain rate relations at rises, faults, and trenches. Such a model has many possibilities to account for past plate motions; hotspots may come and go, and plate migration may radically change the plate-to-plate interactions. But the hotspots would leave visible markers of their past activity on the seafloor and on continents.

❹ This model is compatible with the observation that there is a difference between oceanic island and **oceanic ridge basalts**. It suggests a definite chain of events to form the island type basalt found on Hawaii and parts of Iceland. Relatively primordial material from deep in the mantle rises **adiabatically** up to asthenosphere depths. This partially fractionates into a liquid and solid residual, the liquid rising through **vents** to form the **tholeiitic** part of the island. The latter **alkaline** "cap rocks" would be generated in the lithosphere vent after plate motion had displaced the vent from the "pipe to the deep mantle". In contrast, the ridge basalts would come entirely from the asthenosphere, passively rising to fill the **void** created as plates are pulled apart by the stresses acting on them. The differences in **potassium** and in rare earth pattern for island type and ridge type basalts may be explained by this model. Moreover, the 2-billion-year "holding age" advocated by Gast to explain lead **isotope** data of Gough, Tristan da Cunha, Saint Helena, and Ascension Islands may reflect how long the material was stored in the lower mantle without change prior to the hotspot activity.

❺ My claim that the hotspots provide the driving force for plate motions is based on the following observations. 1) Almost all of the hotspots are near rise crests and there is a hotspot near each of the ridge triple junctions, agreeing with the notion that asthenosphere currents are pushing the plates away from the rises. 2) There is evidence that hotspots become active before continents split apart. 3) The gravity pattern and regionally high **topography** around each hotspot suggest that more than just surface volcanism is involved at each hotspot. 4) Neither rises nor trenches seem capable of driving the plates.

❻ The symmetric magnetic pattern and the "mid-ocean" position of the rises indicate that the rises are passive. If two plates are pulled apart, they split along some line of weakness and in response asthenosphere rises to fill the void. With further pulling of the plates, the laws of heat conduction and the temperature dependence of strength dictate that future cracks appear down the center of the previous "**dike**" injection. If the two plates are displaced equally in opposite directions or if only one plate is moved, and the other held fixed, perfect symmetry of the magnetic pattern will be generated. The axis of the ridge must be

oceanic ridge basalt 洋脊玄武岩
adiabatically *adv.* 绝热地
vent *n.* 通风口,火山口,(空气、气体、液体的)出口
tholeiitic *adj.* 拉斑玄武岩的
alkaline *adj.* 碱性的,含碱的
void *n.* 孔隙
potassium *n.* (化学元素)钾
isotope *n.* 同位素
topography *n.* 地形学,地貌学
dike *n.* 堤坝,沟

spherical harmonics 球(谐)函数
gravity high 重力高
rift n. 裂缝,裂口,地堑,裂谷
v. 断裂,分裂

free to migrate (as shown by the near closure of rises around Africa and Antarctica). If the "dikes" on the ridge axis are required to push the plates apart, it is not clear how the symmetric character of the rises could be maintained. The best argument against the sinking lithospheric plates providing the main motive force is that small trench-bounded plates such as the Cocos Plate[3] do not move faster than the large Pacific Plate. Also, the slow compressive systems would not appear to have the ability to pull other plates away from other units. The pull of the sinking plate is needed to explain the gravity minimum and topographic deep locally associated with the trench system, but I do not wish to invoke this pull as the principal tectonic stress. This leaves sub-lithospheric currents in the mantle and the question now is: are these currents great rolls (mirrors of the rise and trench systems), or are they localized upwellings (that is, hotspots)?

❼ A recent world gravity map computed for **spherical harmonics** up to order 16 shows isolated **gravity highs** over Iceland, Hawaii, and most of the other hotspots. Such gravity highs are symptomatic of rising currents in the mantle. Even if the gravity measurements are inaccurate (different authors have very different gravity maps), the fact remains that the hotspots are associated with abnormally shallow parts of the oceans. For example, note the depth of the million square kilometers surrounding the Iceland, Juan de Fuca, Galapagos, and Prince Edward hotspots. The magnitude of the gravity and topographic effect should measure the size of the mantle flow at each hotspot.

❽ There is evidence of continental expression of hotspot activity in the lands bordering the Atlantic: the Jurassic volcanics in Patagonia (formed by the present day Bouvet Island Plume), the ring dike complex of South-West Africa and flood basalts in the Parana Basin (Tristan da Cunha Plume), the White Mountain Magma series in New Hampshire (the same hotspot that made the New England Seamount Chain (Azores Plume?), the Skaegaard and the Scottish Tertiary Volcanic Province (Iceland Plume), and perhaps others. I claim this line-up of hotspots produced currents in the asthenosphere which caused the continental break-up leading to the formation of the Atlantic. Likewise, the Deccan Traps[4] (Reunion Plume) were symptomatic of the forthcoming Indian Ocean **rifting**. A search should be made for such continental activity, particularly in East Africa and the western United States (the Snake River Basalts?) as an explanation for the rift features found there.

Notes

(1) **The Walvis, Iceland-Faroe** is one island in the Faroe Islands, a group of islands in the North Atlantic Ocean.

(2) **Hawaiian-Emperor, Tuamotu-Line, and Austral-Gilbert-Marshall island chains** are three underwater seamounts. All three chains are approximately parallel and could all have been formed by the same motion of the Pacific Plate over three fixed hotspots.

(3) **The Cocos Plate**, also called the Plate of Coconuts, is a relatively small oceanic plate located the west of Mexico, in the Caribbean. The plate is adjacent to the North American, Caribbean, Rivera, and Pacific plates.

(4) **The Deccan Traps**, located in India and consisting of a composite thickness of more than 2,000 meters of flat basalt lava flows, are one of the largest volcanic provinces in the world.

Reading Comprehension

  Keys

Directions Answer the following questions according to Passage C.

1. According to Paragraph 1, what's the original purpose of the concept of mantle hotspots?

2. How would you define a mantle plume according to Paragraph 2?

3. According to Paragraph 3, why are the existing rises, faults, and trenches everlasting?

4. What are some features of the island-type basalts? How are they different from the oceanic ridge basalts?

5. In Paragraph 5, what evidences support the author's claim that the hotspots provide the driving force for plate motions?

6. How will the perfect symmetry of the magnetic pattern be generated according to Paragraph 6?

7. What do the gravity highs over the hotspots reflect?

8. How does the author explain the rift features of the Atlantic in Paragraph 8?

Exercises *Keys*

Vocabulary

I. Technical words

Directions Match the technical terms with their definitions.

A. biosphere   a. the upper layer of the Earth's mantle, in which there is relatively low resistance to plastic flow, and convection is thought to occur

B. asthenosphere   b. a proposed geological epoch dating from the commencement of significant human impact on the Earth's geology and ecosystems, including, but not limited to, anthropogenic climate change

C. glaciology   c. an upwelling of abnormally hot rock within the Earth's mantle

D. mantle plume   d. a narrow zone on the surface of the Earth where soil, water, and air combine to sustain life (Life can only occur in this zone.)

E. ozone depletion   e. the scientific study of glaciers, or more generally ice and natural phenomena that involve ice

F. aseismic ridge   f. the solid outer section of the Earth which includes the Earth's crust (the outermost layer of rock on the Earth), as well as the underlying cool, dense, and fairly rigid upper part of the upper mantle.

Unit 1 Geoscience

G. Anthropocene   g. the gradual thinning of the Earth's ozone layer in the upper atmosphere caused by the release of chemical compounds containing gaseous chlorine or bromine from industry and other human activities

H. lithosphere   h. a long, linear, and mountainous structure that crosses the basin floor of some oceans, where earthquakes do not occur

II. Academic vocabulary

Directions Fill in the blanks with appropriate forms of the verbs given below. Each word can be used only once.

Verb	Definition in Academic Writing
advance	to put forward an idea, a theory or suggestion
advocate	to support something publicly
assume	to think or accept that something is true without having proof of it
claim	to say that something is true although it has not been proved
construct	to form something by putting different things together
dictate	to control or influence how something happens
generate	to cause something to exist
indicate	to show, point, or make clear in another way
propose	to put forward an idea or plan for consideration or discussion
reflect	to be a sign of the nature of something or of somebody's attitude or feeling

1. Overall, the study _____ that there is no real danger (other than a lack of sleep) to drinking three cups of coffee per day.
2. The plate model was originally _____ to explain the roughly constant surface heat flow at old seafloor.
3. The crystal shape of a mineral _____ the way in which its atoms are arranged.
4. The author, in his latest article, _____ a new theory to explain changes in the climate.
5. Heart specialists strongly _____ low-cholesterol diets.
6. Would you _____ that all plants could survive the winter?
7. Most electricity in the United States is _____ in power plants that use fossil fuels.
8. It was _____ that some doctors were working 80 hours a week.
9. The social conventions of the past _____ that women should remain at home and raise children.
10. For these experiments it is necessary to _____ a model using data from other sources.

Sentence Structure

Understand the structure in English sentences.

There are four types of sentence structure. They are:
- Simple sentences.
- Compound sentences.
- Complex sentences.
- Compound-complex sentences.

A simple sentence has one independent clause, which means only one subject and one predicate.

> For example,
> *The Earth goes round the Sun.*

A compound sentence has at least two independent clauses that are joined together with a coordinating conjunction, like *and*, *but*, *so*, etc.

> For example,
> *It rained for three days, so the streets in my neighborhood flooded.*

A complex sentence has at least one independent and one dependent clause that are joined by subordinating conjunctions, like *because*, *although*, *that*, etc.

> For example,
> *We missed our plane because we were late.*

A compound-complex sentence is a sentence with two or more independent clauses and at least one dependent clause.

> For example,
> *He left in a hurry after he got a phone call but he came back five minutes later.*

Good academic writing is marked by a variety of sentence lengths and structures. Some long sentences can be confusing for readers. No matter how long the sentence is, it belongs to one of the four types. Therefore, analyzing sentence structure and finding out the independent clause(s) will help you understand the long sentence.

Directions Analyze the following sentences by: 1) underlining the independent clause(s); 2) figuring out the sentence type.

1. Visual images of the Earth, in particular *The Blue Marble* image taken by the crew of the Apollo 17 spacecraft on 7 December 1972, sharpened the research focus on the planet as a whole and highlighted its vulnerability to the general public. (Paragraph 4, Passage A)

2. Amidst these developments, J. Lovelock introduced the term "Gaia" in 1972 as an entity comprised of the total ensemble of living beings and the environment with which they interact and hypothesized that living beings regulate the global environment by generating homeostatic feedbacks. (Paragraph 5, Passage A)

3. The scientific developments up to 1980—from Vernadsky's pioneering research, through large-scale field campaigns and the emerging environmental awareness of the 1970s, to Lovelock's Gaia—led to a new understanding of the Earth, challenging a purely geophysical conception of the planet and transforming our view of the environment and nature. (Paragraph 6, Passage A)
4. Although universities maintained their traditional discipline-based faculties, as the emphasis on interdisciplinarity and global-level studies grew, interdisciplinary ESS programs also emerged in many universities around the world. (Paragraph 17, Passage A)

Translation

I. Translate the following English sentences into Chinese. Pay attention to the sentence structure in each sentence.

1. Visual images of the Earth, in particular *The Blue Marble* image taken by the crew of the Apollo 17 spacecraft on 7 December 1972, sharpened the research focus on the planet as a whole and highlighted its vulnerability to the general public.

2. Amidst these developments, J. Lovelock introduced the term "Gaia" in 1972 as an entity comprised of the total ensemble of living beings and the environment with which they interact and hypothesized that living beings regulate the global environment by generating homeostatic feedbacks.

3. The scientific developments up to 1980—from Vernadsky's pioneering research, through large-scale field campaigns and the emerging environmental awareness of the 1970s, to Lovelock's Gaia—led to a new understanding of the Earth, challenging a purely geophysical conception of the planet and transforming our view of the environment and nature.

4. Although universities maintained their traditional discipline-based faculties, as the emphasis on interdisciplinarity and global-level studies grew, interdisciplinary ESS programs also emerged in many universities around the world.

Ⅱ. Translate the following Chinese paragraph into English.

当地球内部发生地震的时候,就会有地震波向四周扩散,这种波可以用地震仪来探测和跟踪,从而锁定地震源的位置。一旦锁定地震源,科学家就可以计算出地震波在地球内部的传播速度。一般来说,地球内部物质构造不同,地震波穿行的速度也会不同。根据穿行速度生成的地球内部图像,我们就可以看到以前从未看过的东西。

Unit 2 Geomaterials

Geomaterials are materials inspired by geological systems originating from the long history of the Earth and may include rock, clay, soils, and so on. They are generally considered as the foundational building blocks of the Geosphere, that is, the minerals and rocks that compose the solid Earth. In this unit, you will read about: 1) how minerals are different from rocks, 2) the properties of minerals, and 3) the attributes of geomaterials used for construction.

Passage A  *Scan and read along the passage*

Rock Forming Minerals*

 *Scan and read along the vocabulary*

❶ The term "minerals" used in a geological sense is not the same as what "mineral" in nutrition labels and **pharmaceutical** products means. In geology, the classic definition of a mineral is a substance that is: 1) naturally occurring, 2) inorganic, 3) solid at room temperature, 4) has an orderly and repeating internal crystalline structure, and 5) a chemical composition that can be defined by a chemical formula. Some natural substances technically should not be considered minerals, but are included by exception. For example, water and mercury are liquid at room temperature. Both are considered minerals because they were classified before the room-temperature rule was accepted as part of the definition. Although the mineral **calcite**, with the chemical formula $CaCO_3$, is quite often formed by organic processes, it is considered a mineral because it is widely found and geologically

pharmaceutical *adj.* 制药的，药品的

calcite *n.* 方解石

---

* This passage is adapted from AFFOLTER M, BENTLEY C, JAYE S, et al., 2020. Earth materials—the rock forming minerals[M/OL]//AFFOLTER M, BENTLEY C, JAYE S, et al. Historical geology. (2020-01-17)[2022-02-10]. https://opengeology.org/historicalgeology/earth-materials/.

discrepancy *n.* 差异,不符
amend *v.* 修改,修订
amber *n.* 琥珀
opal *n.* 欧泊,蛋白石
obsidian *n.* 黑曜石
mineraloid *n.* 准矿石
igneous *adj.* 火成的
sedimentary *adj.* 沉积而成的
gravel *n.* 碎石,沙砾
precipitate *v.* (使)沉淀
solution *n.* 溶解过程
metamorphic *adj.* 变质的
identification *n.* 辨认,识别
outcrop *n.* 露头,露出地面的岩层
erupt *v.* 喷发
magma chamber 岩浆房,岩浆库
fascinating *adj.* 极有吸引力的,迷人的
dub *v.* 把……称为

important. Because of these **discrepancies**, in 1985, the International Mineralogical Association[1] **amended** the definition to a mineral as "A mineral is an element or chemical compound that is normally crystalline and that has been formed as a result of geological processes." Typically, substances like **amber**, pearl, **opal**, or **obsidian** do not fit the definition of mineral because they do not have a crystalline structure. They are referred to as "**mineraloids**".

❷ In contrast to minerals, rocks are more loosely defined. Simply, a rock is a solid substance that is made of one or more minerals or mineraloids. There are three families of rocks: **igneous** (rocks crystallizing from molten materials), **sedimentary** (rocks composed of the products of mechanical weathering, like sand, **gravel**, etc., and/or chemical weathering, like minerals and mineraloids **precipitated** from **solution**), and **metamorphic** (rocks produced by the chemical and physical reorganization of other rocks under conditions induced by elevated heat and/or pressure).

❸ Mineral **identification** is the first step in understanding the formation of a rock and its history. Geologists learn to "read the rock" to understand the Earth's history at any given location where a rock is found in an **outcrop**. This allows geologists to understand what the environment was like at the moment the rock formed. Was there a volcano **erupting** or does the rock tell us that it formed deep inside a **magma chamber**? Was the rock formed by burial of an ancient beach? Was the rock formed by compressive forces deep within the crust as continents collided and new mountains were forming? The clues to these widely different environments of formation are "written in the rocks". The first step in understanding the rocks' history is being able to identify, characterize, and quantify the minerals that compose the rocks. Rocks are fascinating to a geologist because every rock has a story to tell. As they read the rock from one location to the next, it helps them piece together the **fascinating** story of the Earth.

❹ As of 2018, there were over 5,000 minerals officially recognized by the International Mineralogical Association. Many of these formed under very specific chemical and geological conditions and may only occur in one location on the Earth. Fortunately for geology students, only a small subset of these minerals is common and thus it is truly necessary to identify the Earth's most common rocks. These minerals have been **dubbed** "the Big Ten" by the prominent American mineralogist Mickey Gunter in *Mineralogy and Optical Mineralogy* [2]. The ten minerals are **olivine, augite, hornblende, biotite, anorthite** (**calcium**-rich **plagioclase**), **albite** (**sodium**-rich plagioclase), **orthoclase** (potassium-rich **feldspar**), **muscovite, quartz**, and calcite.

❺ Minerals form when atoms bond together in a crystalline arrangement. In order for a mineral crystal to grow, the elements needed to make it must be present in the appropriate proportions, the physical and chemical conditions must be favorable, and there must be sufficient time for the atoms to become arranged.

❻ Physical and chemical conditions include factors such as temperature, pressure, presence of water, pH, and the amount of oxygen available. Time is one of the most important factors because it takes time for atoms to move from place to place and become ordered. If time is limited, the mineral grains (with crystalline structure) will remain very small. The presence of water enhances the mobility of **ions** and can lead to the formation of larger crystals over shorter time periods.

❼ The Earth's crust is dominantly composed of just a few mineral groups, the vast majority of them are **silicate** minerals. The following processes are responsible for the formation of the various mineral groups that compose the crust.

- Crystallization from molten rock material (magma or **lava**). The majority of minerals in the crust have formed this way.
- Organic formation: formation of minerals by organisms within shells (primarily calcite), teeth, and bones (primarily **apatite**)
- **Precipitation** from **aqueous** solution (**i. e.**, from hot water flowing underground, from evaporation of a lake or inland sea, or in some cases, directly from seawater).
- Weathering: during which minerals unstable at the Earth's surface (conditions of low temperature, low pressure, high moisture, and high oxygen levels) may be altered to other minerals.
- Metamorphism: formation of new minerals directly from the elements within existing minerals under conditions of elevated temperature and/or pressure.

❽ Most of the minerals of the Earth's crust formed through the cooling of molten rock (magma or lava). Molten rock is very hot, typically on the order of 1,000 ℃ (1,800 F) or more. Heat is

olivine *n.* 橄榄石,黄绿
augite *n.* 辉石
hornblende *n.* 角闪石
biotite *n.* 黑云母
anorthite *n.* 钙长石
calcium *n.* (化学元素)钙
plagioclase *n.* 斜长石
albite *n.* 钠长石
sodium *n.* 钠
orthoclase *n.* 正长石
feldspar *n.* 长石
muscovite *n.* 白云母
quartz *n.* 石英
ion *n.* 离子
silicate *n.* 硅酸盐
lava *n.* (火山)熔岩,岩浆
apatite *n.* 磷灰石
precipitation *n.* 降水(如雨、雪、冰雹)
aqueous *adj.* 水的,水般的
i. e. (id est) 即,也就是

vibrate *v.* (使)震动,(使)颤动
concentration *n.* 含量,浓度
occurrence *n.* 出现,发生
accumulation *n.* 积聚

energy and temperature is a measure of that energy. Heat causes atoms to **vibrate**, and temperature measures the intensity of the vibration. If vibrations are very strong, chemical bonds will break, the pre-existing minerals in the Earth's crust, and mantle will melt. Melting releases ions into a pool, forming a magma chamber. Magma is simply molten rock with freely moving ions. If magma is allowed to cool at depth or erupted onto the surface (then called lava), mineral crystals will form as temperature decreases.

❾ The chemical composition of the Earth's crust is critical to the discussion of the dominant minerals to form from magma or lava, the silicate minerals. It is important to understand that not every location in the crust will have this composition. Rocks are different from one to another, and virtually no two rocks within the crust will have the exact same composition. The same also goes for magma. No two magma bodies will have the exact same composition. The magma will, however, contain these elements in somewhat varying proportions depending on where exactly within the crust (or mantle) the magma was derived. What we can say is that the magma will be largely composed of silicon and oxygen with varying proportions of the remaining six elements plus other elements in trace amounts. The combination of these elements will form the different silicate minerals as the magma cools either deep within the Earth or near/at the surface.

❿ Over the years, geologists have been keenly interested in learning how natural processes produce localized **concentrations** of essential minerals. One well-established fact is that **occurrences** of valuable mineral resources are closely related to the rock cycle. That is, the mechanisms that generate igneous, sedimentary, and metamorphic rocks, including the processes of weathering and erosion, play a major role in producing concentrated **accumulations** of useful elements.

Notes

(1) **International Mineralogical Association (IMA)**, founded in 1958, is the world's largest organization promoting mineralogy, one of the oldest branches of science. 38 national mineralogical societies or groups are members of IMA. The Association supports the

activities of Commissions and Working Groups involved on certain aspects of mineralogical practice and facilitates interaction among mineralogists by sponsoring and organizing meetings.

(2) ***Mineralogy and Optical Mineralogy*** is a college-level textbook designed for courses in rocks and minerals, mineralogy, and optical mineralogy. It covers crystallography, crystal chemistry, systematic mineralogy, and optical mineralogy. The textbook is organized to facilitate spiral learning, with introductory through advanced chapters on each of these four topics. A chapter on hand sample identification is also included.

Reading Comprehension

 Keys

Directions Answer the following questions according to Passage A.

1. Do you think water is a mineral? Why or why not?

2. Why is amber referred to as a mineraloid rather than a mineral?

3. How are rocks different from minerals? How are they related to each other?

4. Why and how do geologists "read the rock"?

5. What are the criteria for selecting the ten minerals?

6. How were the minerals in the crust formed?

7. How do you understand "virtually no two rocks within the crust will have the exact same composition"?

8. Sort the given words into appropriate categories.

①water　②mercury　③olivine　④augite　⑤hornblende　⑥amber　⑦pearl　⑧opal　⑨obsidian　⑩biotite　⑪anorthite　⑫albite　⑬orthoclase　⑭muscovite　⑮quartz　⑯calcite　⑰igneous　⑱sedimentary　⑲metamorphic　⑳apatite

Minerals	
Mineraloid	
The Big Ten	
Rocks	

Passage B

Scan and read along the passage

Scan and read along the vocabulary

sophisticated *adj.* 复杂巧妙的,先进的,精密的
commodity *n.* 商品,货物
external *adj.* 外部的,外面的
internal *adj.* 内部的,体内的
restriction *n.* (受)限制(状态),(受)约束(状态)
distinctive *adj.* 独特的,与众不同的

Physical Properties of Mineral*

❶ A mineral is a naturally occurring solid formed by inorganic processes. Since the internal structure and chemical composition of a mineral are difficult to determine without the aid of **sophisticated** tests and apparatus, the more easily recognized physical properties are used in identification.

❷ Most people think of a crystal as a rare **commodity**, when in fact most inorganic solid objects are composed of crystals. The reason for this misconception is that most crystals do not exhibit their crystal form: the **external** form of a mineral that reflects the orderly **internal** arrangement of its atoms. Whenever a mineral forms without space **restrictions**, individual crystals with well-formed crystal faces will develop. Some crystals, such as those of the mineral quartz, have a very **distinctive** crystal form that can be helpful in identification. However, most of the time, crystal growth is interrupted because of competition for space, resulting

---

* This passage is adapted from 知乎专栏,2022.托福阅读 TPO61-1:Physical properties of minerals[EB/OL].[2022-04-22]. https://zhuanlan.zhihu.com/p/441833554.

in an **intergrown** mass of crystals, none of which exhibits crystal form.

❸ Although color is an obvious feature of a mineral, it is often an **unreliable diagnostic** property. Slight **impurities** in the common mineral quartz, for example, give it a variety of colors, including pink, purple (**amethyst**), white, and even black. When a mineral, such as quartz, exhibits a variety of colors, it is said to possess **exotic** coloration. Exotic coloration is usually caused by the inclusion of impurities, such as foreign ions, in the crystalline structure. Other minerals for example, sulfur, which is yellow, and **malachite**, which is bright green-are said to have inherent coloration because their color is a consequence of their chemical makeup and does not vary significantly.

❹ **Streak** is the color of a mineral in its powdered form and is obtained by rubbing a mineral across a plate of **unglazed porcelain**. Whereas the color of a mineral often varies from sample to sample, the streak usually does not and is therefore the more reliable property.

❺ **Luster** is the appearance or quality of light reflected from the surface of a mineral. Minerals that have the appearance of metals, regardless of color, are said to have a metallic luster. Minerals with a nonmetallic luster are described by various adjectives, including **vitreous** (glassy), pearly, silky, **resinous**, and earthy (dull).

❻ One of the most useful diagnostic properties of a mineral is hardness, the resistance of a mineral to **abrasion** or **scratching**. This property is determined by rubbing a mineral of unknown hardness against one of known hardness, or vice versa. A numerical value can be obtained by using Mohs Scale of Hardness[1], which consists of ten minerals arranged in order from **talc**, the softest, at number one, to diamond, the hardest, at number ten. Any mineral of unknown hardness can be compared with these or with other objects of known hardness. For example, a fingernail has a hardness of 2.5, a copper penny 5, and a piece of glass 5.5. The mineral **gypsum**, which has a hardness of two, can be easily scratched with your fingernail. On the other hand, the mineral calcite, which has a hardness of three, will scratch your fingernail but will not scratch glass quartz, the hardest of the common minerals, will scratch a glass plate.

intergrown *adj.* 共生的
unreliable *adj.* 不可靠的,靠不住的
diagnostic *adj.* 判断的
impurity *n.* 杂质
amethyst *n.* 紫水晶
exotic *adj.* 奇异的,异国风情的
malachite *n.* 孔雀石
streak *n.* 条纹,条痕
unglazed *adj.* 未上釉的
porcelain *n.* 瓷,瓷器
luster *n.* 光泽,光彩
vitreous *adj.* 玻璃状的,透明的
resinous *adj.* 树脂质的,像树脂的
abrasion *n.* 磨损,擦伤
scratching *n.* 擦伤,刮痕
talc *n.* 滑石
gypsum *n.* 石膏

cleavage *n.* 解理
mica *n.* 云母
configuration *n.* 布局,构造,配置
fracture *v.* 破裂,折断,瓦解,分裂
curve *v.* (使)沿曲线移动,呈曲线形
resemble *v.* 像,与……相似
splinter *n.* 碎片,微小的东西
fiber *n.* 纤维

❼ The tendency of a mineral to break along planes of weak bonding is called **cleavage**. Minerals that possess cleavage are identified by the smooth, flat surfaces produced when the mineral is broken. The simplest type of cleavage is exhibited by the **micas**. Because the micas have excellent cleavage in one direction, they break to form thin, flat sheets. Some minerals have several cleavage planes, which produce smooth surfaces when broken, while others exhibit poor cleavage, and still others exhibit no cleavage at all. When minerals break evenly in more than one direction, cleavage is described by the number of planes exhibited and the angles at which they meet. Cleavage should not be confused with crystal form. When a mineral exhibits cleavage, it will break into pieces that have the same **configuration** as the original sample does. By contrast, quartz crystals do not have cleavage, and if broken, would shatter into shapes that do not resemble each other or the original crystals. Minerals that do not exhibit cleavage are said to **fracture** when broken. Some break into pieces with smooth **curved** surfaces **resembling** broken glass. Others break into **splinters** or **fibers**, but most fracture irregularly.

Note

(1) **Mohs Scale of Hardness** is used as a convenient way to help identify minerals. A mineral's hardness is a measure of its relative resistance to scratching, measured by scratching the mineral against another substance of known hardness on the Mohs Hardness Scale, as shown in the following table.

	Mineral Name	Scale	Common Object
increasing hardness ↓	talc	1	
	gypsum	2	手指甲(fingernail, 2.5)
	calcite	3	
	fluorite	4	水果刀(knife, 5.5)
	apatite	5	
	feldspar	6	钢钉(steel nail, 6.5)
	quartz	7	
	topaz	8	钻头(drill bit, 8.5)
	corundum	9	
	diamond	10	

Reading Comprehension

 Keys

Directions Answer the following questions according to Passage B.

1. The word "apparatus" in Paragraph 1 is closest in meaning to
   A. equipment.
   B. procedures.
   C. experiments.
   D. laboratories.

2. According to Paragraph 2, which of the following is a mistaken belief that people have about crystals?
   A. Crystals always have a well-formed shape.
   B. Minerals are generally composed of crystals.
   C. The atoms of a crystal have an orderly arrangement.
   D. Crystals are always solid and inorganic.

3. According to Paragraph 3, how do different samples of the same mineral come to exhibit a variety of colors?
   A. The samples have different crystalline structures.
   B. The samples contain different varieties of quartz.
   C. The samples differ in the impurities they contain.
   D. The samples were formed in different exotic conditions.

4. The word "inherent" in Paragraph 3 is closest in meaning to
   A. bright.
   B. essential.
   C. superficial.
   D. transparent.

5. Which of the following can be inferred about streak from Paragraph 4?
   A. When a sample of a mineral is rubbed on unglazed porcelain, the color of the streak is usually the same as the color of the sample.
   B. In most cases, different samples of a mineral produce streaks that are all of the same color even though the samples themselves are of different colors.
   C. When a streak is made, the impurities in the mineral are removed, and the true color of the mineral is revealed.
   D. Streak color is reliable for identifying minerals because a given mineral sample always yields the same color of streak each time it is rubbed.

6. It can be inferred from Paragraph 6 that the mineral quartz
   A. has no fixed degree of hardness.
   B. might scratch the surface of a diamond.

C. is harder than calcite.

D. has atoms that are weakly bonded to each other.

7. According to Paragraph 7, which of the following is true of the cleavage of micas?

   A. Micas are the only minerals to break along planes of weak bonding.

   B. Micas exhibit poor cleavage because they tend to break unevenly.

   C. Micas break in one direction, forming thin sheets with smooth surfaces.

   D. Micas break in several directions, forming a number of angles and planes.

8. Why does the author warn that cleavage should not be confused with crystal form?

   A. Because most people have the mistaken belief that the surfaces of crystals are planes of crystal cleavage.

   B. Because the author's characterization of cleavage in terms of smooth planes and the angles between them could easily be mistaken for a description of crystal form.

   C. To make the point that crystal form and cleavage are the same property only in the simplest cases of cleavage, such as mica.

   D. To introduce a discussion of minerals the have cleavage but not crystal form.

9. The letters [A], [B], [C], and [D] indicate where the following sentence (in bold) could be added to the following part from the passage. Where would the sentence best fit?

   Each mineral has an orderly arrangement of atoms (crystalline structure) and a definite chemical composition that give it a unique set of physical properties.

   A mineral is a naturally occurring solid formed by inorganic processes. [A] Since the internal structure and chemical composition of a mineral are difficult to determine without the aid of sophisticated tests and apparatus, the more easily recognized physical properties are used in identification.

   [B] Most people think of a crystal as a rare commodity, when in fact most inorganic solid objects are composed of crystals. [C] The reason for this misconception is that most crystals do not exhibit their crystal form: the external form of a mineral that reflects the orderly internal arrangement of its atoms. [D] Whenever a mineral forms without space restrictions, individual crystals with well-formed crystal faces will develop. Some crystals, such as those of the mineral quartz, have a very distinctive crystal form that can be helpful in identification. However, most of the time, crystal growth is interrupted because of competition for space, resulting in an intergrown mass of crystals, none of which exhibits crystal form.

10. An introductory sentence (in bold) for a brief summary of the passage is provided below. Complete the summary by selecting the THREE answer choices that express the most important ideas in the passage. ( )

    Minerals have a number of physical properties, some of which are useful for mineral identification.

A. Most minerals show a characteristic crystal form that results from the way their atoms are arranged in the crystal structure.

B. Luster is a good guide to the identity of glassy and metallic minerals, but not those that have dull or earthy appearance.

C. Some minerals characteristically split along one or more smooth planes of cleavage while others typically fracture unevenly.

D. For many minerals, the streak formed by a rubbing a mineral sample on unglazed porcelain is a more reliable guide for identification that the color of the sample.

E. The relative hardness of a mineral as determined by Mohs scale is very useful for identifying a mineral sample.

F. Quartz has a characteristic crystal form, but no cleavage and its color varies widely from sample to sample, all of which make it a typical mineral.

Passage C

Attributes of Geomaterials in Construction*

❶ Traditionally, the occurrence of a specific material close to **dwellings** was significant in terms of the energy required to transport the material once **extracted**. For ordinary structures, the **hauling** distance was the most important **parameter**, and often a material occurring nearby, even of lower quality, was preferred to a more **durable** one that had to be transported over several tens of kilometers. Therefore, the use of very local resources of raw materials for construction underpins the appearance of historical urban/village areas and can be considered an important sustainable approach allowing decreased impact of raw material consumption at least from the perspective of reducing negative impacts from long-distance transport. Transport of the material over larger distances was practiced only if local geological conditions did not provide suitable material or if a material of certain properties, such as aesthetic qualities, was sought. Even in very recent times, most **aggregates** can be transported economically over a relatively short distance.

attribute n. 属性,特质
dwelling n. 住宅,住所
extract v. 取出,拔出,提取,提炼
hauling n. 搬运,拖运
parameter n. 参数,变量
durable adj. 持久的,耐用的
aggregates n. 混凝土

---

* This passage is adapted from PŘIKRYL R, TÖRÖK Á, GÓMEZ-HERAS M, et al., 2016. Sustainable use of traditional geomaterials in construction practice[M]. London: Geological Society of London.

forego *v.* 发生在……之前
dimension *n.* 大小，尺寸
blocky stone 块状石头
dimension stone 规格石材
archaeological *n.* 考古学的，考古的
embody *v.* 使具体化
pillar *n.* 柱子，支柱
millennia *n.* 千年
binder *n.* 黏合剂
pyrotechnology *n.* 高温技术

❷ Once the material was identified as accessible at the site, it had to be extracted and processed in a certain manner. The processing of geomaterials into structural elements or functional products that can be employed in a built environment presents one of the least explored features of the history of humankind. At least in the case of natural stone, these skills were most probably acquired based on previous experience of extraction and manufacturing of harder rocks used for chipping stone tools and/or weapons. However, the range of materials used in construction and the variety of processing methods also mean that many new "discoveries" had to appear in a period **foregoing** the invention of pottery firing and/or metal smelting.

❸ The first important "invention" related to inorganic construction materials was the discovery of the possibility of extracting larger pieces—blocks of certain **dimensions**; connection of "**blocky stone**" and "dimension" led to the denomination of "natural stone" as "**dimension stone**" more recently. These rough blocks of natural stone could be further processed by cutting or carving in order to receive a desired shape. In the most prominent cases, the final 3D product was not only cut/carved to attain certain dimensions and/or dressed on the surface, but also decorated with carved 3D reliefs, or the whole block was sculpted. Interestingly, this invention is first documented on a larger scale from Göbekli Tepe[1], an archaeological site located in today's southern Turkey. According to a detailed **archaeological** survey of the site, the development of numerous circular structures with **embodied** T-shaped **pillars** from local natural stone occurred from the tenth to the eighth **millennia** BCE[2], that is, at a very beginning of the Holocene[3].

❹ Another important "innovation" is linked to the discovery of methods of producing and utilizing inorganic **binders**, of which air lime was the first. Importantly, the first documented uses of air lime as a burnt material prepared by the intentional burning of a natural raw material happened several thousand years before the firing of pottery and/or the smelting of ores for metals. The burning of lime and its use in construction is of the upmost importance. In a broader context of acquiring technological skills, it can be acknowledged as the beginning of **pyrotechnology**, that is, the intentional use of fire for the production of new materials from natural ones. From the material point of view, the burning of natural materials into a new material was an extraordinarily important invention because it means that early man had to recognize that the original material had undergone a kind of transformation by the means of fire. From this recognition and the **subsequent** processing, the material gained new properties that made it suitable for binding **granular** elements in **mortar**, for binding together larger constructional elements or even for painting walls.

❺ The above historic context shows that if a certain material is to be used in construction, it must be: 1) easily available, 2) workable, and 3) serviceable. The first of these attributes means that, in the case of inorganic materials used for constructional purposes, the raw materials are derived from the Earth, mostly from its uppermost parts known as the Earth's crust, or in some case the very thin layer that results from the interactions between lithosphere, atmosphere, and biosphere that is termed the **pedosphere**, from which soils for construction are dug. Only if the material is available in sufficient quality and quantity from the surface (open-air operation—quarrying) or near the surface (underground operation—mining), can it be claimed to be available.

❻ When the material is available for extraction, and the extraction is allowed at specific sites, the second of three fundamental attributes of construction materials must exist, specifically the knowledge of how to extract it and how to process the extracted raw material to obtain the desired properties before its use in construction. Whilst the accessibility depends mainly on the Earth's processes plus on humankind's ability to find the material and to extract it, the workability is a reflection of man's ability to understand the properties of the raw material and the knowledge/craftsmanship to process it in order to achieve the desired **composition** and physical properties (i.e., the ability of humans to transform the raw material into a new one by applying certain technologies, such as brick firing or shaping rough blocks of dimension stone into **ashlars, paving/cladding slabs** or even statues). Workability also represents an important aspect of correct conditioning and preparation of materials **prior to** their use.

❼ In contrast to the previous two functional attributes, the

subsequent *adj.* 随后的，接着的
granular *adj.* 颗粒的，粒状的
mortar *n.* 砂浆，灰浆
pedosphere *n.* （地球的）土壤圈
composition *n.* 成分构成，成分
ashlar *n.* 琢石
paving/cladding slab 铺路板
prior to 在……之前

petrographical *adj.* 岩相学的,岩石学的
glory *n.* 荣誉
aesthetic *adj.* 审美的,美学的
withstand *v.* 经受住,承受住
vital *adj.* 至关重要的,必不可少的
minute *adj.* 极小的,微小的,详细的,细致入微的
fatal *adj.* 致命的
destruction *n.* 破坏,摧毁
sanctuary *n.* 宗教圣地,(教堂内的)圣坛
fortification *n.* 碉堡,防御工事
sculpture *n.* 雕像,雕塑作品
longevity *n.* 持续时间,耐用期限

serviceability of materials used in construction is a different issue. In a broad sense, serviceability can be understood as a combination of numerous properties—both physical ones and technical ones—that secure the stability of the structure (e.g., bridge, building) or of an object (e.g., statue) in a certain environment over a desired period of time (the concept of design life). Understanding the complexity of relationships between various **petrographical** characteristics and physical and/or technological properties is important to secure the stability and proper function of the construction and also to provide sufficient durability.

❽ Although serviceability is used here as the more general term, durability is a very close synonym. Often simplified as equal to weathering resistance or technical serviceability, the desired functional property of a supposed durable material, often in the requirement of certain varieties of natural stone used in monuments, was to last forever and thus to transmit the **glory** or the genius of the commander to the followers. To fulfil this function, the material must retain its shape and appearance, thus maintaining the material's structural and **aesthetic** functions. The ability to **withstand** the action of weathering processes is **vital** for artistically carved surfaces where the loss of even **minute** surface layers can result in the **fatal destruction** of the artistic meaning of the whole object. Specifically, for prestigious structures such as **sanctuaries**, palaces, defense **fortifications**, and **sculptures**, the **longevity** of the materials from which these were built or manufactured was, along with aesthetic appearance, the key function. In this sense, durability was and still is considered as the principal functional property of construction materials, although it cannot be directly measured or predicted from any single test procedure.

Notes

(1) **Göbekli Tepe** is an ancient city in southeastern Turkey once named "Edessa" and known as "the City of the Prophets" (预言者;先知). It is believed that the site used to be a holy place of ritual significance. It is marked by layers of carved stone memorials and is estimated to date to the 10th millennium BC. The site is listed by UNESCO as a World Heritage Site.

(2) **BCE** (Before Common Era) and **CE** (Common Era) are used to indicate if a year is before or after the first year of Common Era. Many countries prefer to use the more modern and neutral CE and BCE to replace the traditional abbreviations BC and AD because they hold religious (Christian) meanings.

(3) **Holocene** (全新世) is the geological period extending from the present day back to about 10,000 radiocarbon years, about 11,430 ± 130 calendar years before present (BP) (between 9560 BC and 9300 BC). Holocene is the fourth and last epoch of the Neogene Period (新近纪) of the Cenozoic Era (新生代).

Reading Comprehension

 Keys

Directions Fill in the blanks according to Passage C.

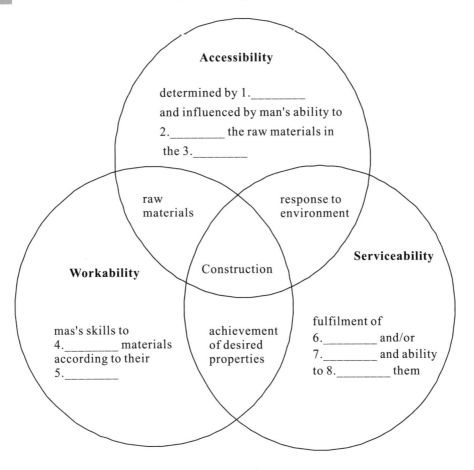

Exercises

 Keys

Vocabulary

Ⅰ. Technical words

Directions Write out the English expressions according to the Chinese.

准矿石　　　　　　　＿＿＿＿＿＿＿＿＿＿＿＿
沉积岩　　　　　　　＿＿＿＿＿＿＿＿＿＿＿＿
岩浆房　　　　　　　＿＿＿＿＿＿＿＿＿＿＿＿
硅酸盐　　　　　　　＿＿＿＿＿＿＿＿＿＿＿＿
规格石材　　　　　　＿＿＿＿＿＿＿＿＿＿＿＿
黏合剂　　　　　　　＿＿＿＿＿＿＿＿＿＿＿＿
土壤圈　　　　　　　＿＿＿＿＿＿＿＿＿＿＿＿
铺路板　　　　　　　＿＿＿＿＿＿＿＿＿＿＿＿

Ⅱ. Academic vocabulary

Directions Fill in the blanks with appropriate forms of the words given below. Each word can be used only once.

substance	critical	enhance	property	durable
occurrence	identification	obtain	classify	consequence

1. Minerals are crystalline _____, made up of atoms (or ions) that are arranged in an orderly, repetitive manner.
2. Geologists _____ igneous rocks on the basis of their texture and mineral composition.
3. The other common mineral, quartz, is abundant because it is extremely _____ and very resistant to chemical weathering.
4. Although dams are effective in reducing flooding and provide other benefits, their construction and maintenance also have significant costs and _____.
5. Some minerals exhibit a variety of colors. Thus, the use of color as a means of _____ is often ambiguous or even misleading.
6. Another optical (光学的) _____ used to identify minerals is the ability to transmit light.
7. One well-established fact is that _____ of valuable mineral resources are closely related to the rock cycle.
8. Groundwater represents the largest reservoir of freshwater that is readily available to humans and is a _____ resource for human civilization.
9. In practice, however, different recording stations often _____ slightly different magnitudes for the same earthquake—a result of the variations in the rock types through which the waves travel.

Unit 2 Geomaterials

10. Crystallization in a fluid-rich environment, where ion migration is _____, results in the formation of unusually large crystals.

Sentence Structure

Understand parenthesis in English sentences.

A parenthesis is a word, phrase, or clause inserted into a sentence as an explanation or afterthought. A parenthesis is usually marked off by brackets, commas, or dashes. When a parenthesis is removed, the surrounding text is still grammatically complete.

> For example,
> I, *as they say*, am as free as a bird.
> George Washington—*the first American president*—was born in 1732.
> It's unusual to see John sleep so early (*as he often studied late into the night*).

In academic writing, parentheses within brackets are most often used to convey technical information such as equations, to introduce acronyms, and for parenthetical citations.

> For example,
> Several prestigious organizations, such as the National Institutes of Health *(NIH)*, supported the initiative.
> This sort of testing, as Jenner *(2012)* said, is usually unreliable.

Directions Analyze the following sentences by: 1) figuring out the independent clause(s), and 2) underlining the parentheses.

1. For ordinary structures, the hauling distance was the most important parameter and often a material occurring nearby, even of lower quality, was preferred to a more durable one that had to be transported over several tens of kilometers. (Paragraph 1, Passage C)
2. Only if the material is available in sufficient quality and quantity from the surface (open-air operation—quarrying) or near the surface (underground operation—mining), can it be claimed to be available. (Paragraph 5, Passage C)
3. In a broad sense, serviceability can be understood as a combination of numerous properties—both physical ones and technical ones—that secure the stability of the structure (e.g., bridge, building) or of an object (e.g., statue) in a certain environment over a desired period of time (the concept of design life). (Paragraph 7, Passage C)
4. Often simplified as equal to weathering resistance or technical serviceability, the desired functional property of a supposed durable material, often in the requirement of certain varieties of natural stone used in monuments, was to last forever and thus to transmit the glory or the genius of the commander to the followers. (Paragraph 8, Passage C)

Translation

Ⅰ. **Translate the following English sentences into Chinese. Pay attention to the parenthesis in each sentence.**

1. For ordinary structures, the hauling distance was the most important parameter and often a material occurring nearby, even of lower quality, was preferred to a more durable one that had to be transported over several tens of kilometers.

2. Only if the material is available in sufficient quality and quantity from the surface (open-air operation—quarrying) or near the surface (underground operation—mining), can it be claimed to be available.

3. In a broad sense, serviceability can be understood as a combination of numerous properties—both physical ones and technical ones—that secure the stability of the structure (e.g., bridge, building) or of an object (e.g., statue) in a certain environment over a desired period of time (the concept of design life).

4. Often simplified as equal to weathering resistance or technical serviceability, the desired functional property of a supposed durable material, often in the requirement of certain varieties of natural stone used in monuments, was to last forever and thus to transmit the glory or the genius of the commander to the followers.

Ⅱ. Translate the following Chinese paragraph into English.

颜色是矿物最明显的特征,几乎所有人鉴别矿物都首先从颜色入手。矿物的颜色可以帮助我们缩小备选范围。但是,大部分的矿物可能都不止有一种颜色,比如萤石、玛瑙、方解石、刚玉。它们由于各种原因,有着非常丰富的色彩种类。而且不同矿物也常有相同的颜色。如果仅凭颜色是不易分辨的,所以还需要参照其他的性质辅助鉴别。

Unit 3　Atmospheric and Climate Sciences

Climate has a profound impact on many geologic processes. When climate changes, these processes respond. To accurately portray the character of a place or an area, climate variations and extremes must also be included. It is becoming clear that climate change is getting severe and that the impact of climate change is variant and vast. Many of the changes will have powerful economic, social, and political consequences. What are important concepts regarding global climate change? How is urban development related to climate? What are implications of a history of ENSO events for future climate change? These are the questions to be answered by the passages of this unit.

Passage A  *Scan and read along the passage*

Global Climate Change*

 *Scan and read along the vocabulary*

❶ Anyone who has the opportunity to travel around the world will find such an incredible **variety** of climates that it is hard to believe they could all occur on the same planet. Climate not only **varies** from place to place, but it is also naturally **variable** over time. Over the great expanse of the Earth history, and long before humans were roaming the planet, there were many shifts—from warm to cold, and from wet to dry, and back again. The **geologic** record is a storehouse of data that confirms this fact.

❷ Today, global climate change is more than just a topic of academic interest to a group of scientists curious about the Earth history. Rather, the subject is making headlines. The reason is that research focused on human activities and their impact on the

variety *n.* 多样化,变化
vary *v.* 变化
variable *adj.* 易变的,多变的
geologic *adj.* 地质的

---

* The passage is excerpted from LUTGENS F K, TARBUCK E J, TASA D, 2011. Essentials of geology[M]. 11th ed. Upper Saddle River, NJ: Pearson Prentice Hall.

inadvertently *adv.* 无意地,不经意地
variation *n.* 变化,变动
variability *n.* 可变性,变化性
multidimensional *adj.* 多维的
hydrosphere *n.* 地球水圈
cryosphere *n.* 冰冻圈
decipher *v.* 破译
proxy data 代用资料,替代性指标
sediment *n.* 沉积物
fossil pollen 花粉化石
paleoclimatology *n.* 古气候学
nitrogen *n.* 氮

environment has demonstrated that people are **inadvertently** changing the climate. Unlike changes in the past, which were natural **variations**, modern climate change is dominated by human influences that are sufficiently large as to exceed the bounds of natural **variability**. Moreover, these changes are likely to continue for many centuries.

❸ The Earth is a **multidimensional** system that consists of many interacting parts. A change in any one part can produce changes in any or all of the other parts. This fact is certainly true when it comes to the study of climate and climate change. There is a climate system that includes the atmosphere, **hydrosphere**, geosphere, biosphere, and **cryosphere**. The climate system involves the exchanges of energy and moisture that occur among the five spheres. These exchanges link the atmosphere to the other spheres so that the whole functions as an extremely complex, interactive unit. When one part of it changes, the other components also react.

❹ High-technology and precision instrumentation are now available to study the composition and dynamics of the atmosphere. Such tools are recent inventions, however, and have been providing data for only a short time span. Instrumental records go back only a couple of centuries at best, and the further back we go, the less complete and more unreliable the data become. To understand fully the behavior of the atmosphere and to anticipate future climate change, scientists **decipher** and reconstruct past climates by using indirect evidence. Such **proxy data** come from natural recorders of climate variability, such as seafloor **sediments**, glacial ice, **fossil pollen**, and tree growth rings. Historical documents can also be useful sources of information about climate variability. The main goal of **paleoclimatology** is to understand the climate of the past in order to assess the current and potential future climate in the context of natural climate variability.

❺ In order to better understand climate change, it is helpful to possess some knowledge about the composition, structure of the atmosphere and the process by which it is heated—the greenhouse effect. Clean, dry air is composed almost entirely of two gases— 78 percent **nitrogen** and 21 percent oxygen. Although these gases are the most plentiful components of air and are of great significance to life on the Earth, they are of little or no importance in affecting weather phenomena. The remaining 1 percent of dry air is mostly the **inert gas argon** (0.93 percent) plus tiny quantities of a number of other gases. Carbon dioxide, although present in only minute amounts (0.038 9 percent), is nevertheless an important **constituent** of air because it has the ability to absorb heat energy radiated by the Earth and thus influences the heating of the atmosphere. Some other important gases and particles in the air include **water vapor**, ozone, and **aerosols**.

❻ The atmosphere rapidly thins as you travel away from the Earth, until there are too few gas **molecules** to detect. Obviously, the atmospheric pressure at higher **altitudes** is less. In addition to vertical changes in air pressure, there are also changes in air temperature as we ascend through the atmosphere. The bottom layer in which we live is characterized by a decrease in temperature with an increase in altitude and is called the **troposphere**. The troposphere is the chief focus of **meteorologists**, because it is in this layer that essentially all important weather phenomena occur.

❼ Nearly all of the energy that drives the Earth's variable weather and climate comes from the Sun. Although these forms of energy comprise a major portion of the total energy that radiates from the Sun, they are only part of a large array of energy called radiation or **electromagnetic** radiation. All radiation transmits energy through the vacuum of space at 300,000 kilometers per second and only slightly slower through our atmosphere. When an object absorbs any form of **radiant** energy, the result is an increase in molecular motion, which causes a corresponding increase in temperature.

❽ On average, about 50 percent of the solar energy reaching the top of the atmosphere passes through the atmosphere and is absorbed at the Earth's surface. Another 20 percent is absorbed directly by clouds and certain **atmospheric gases** (including oxygen and ozone) before reaching the surface. The remaining 30 percent is reflected back to space by the atmosphere, clouds, and reflective surfaces such as snow and ice. The extremely important role the atmosphere plays in heating the Earth's surface has been named the greenhouse effect.

❾ A great variety of **hypotheses** have been proposed to explain climate change. Two hypotheses have earned serious consideration

inert gas 惰性气体
argon n. 氩
constituent n. 成分,构成要素
water vapor 水蒸气
aerosol n. 气溶胶
molecule n. 分子
altitude n. 海拔高度
troposphere n. 对流层
meteorologist n. 气象学者
electromagnetic adj. 电磁的
radiant adj. (热、能量)辐射的
atmospheric gas 大气气体
hypothesis n. 假设(复数为hypotheses)

volcanic *adj.* 火山的，由火山引发的
volcanic eruption 火山喷发
emit *v.* 排放，散发
fine-grained *adj.* 细粒的
debris *n.* 碎片，残骸
solar radiation 太阳辐射
meteorologically *adv.* 从气象学角度看
transparent *adj.* 可穿透的，(热、电磁波)可通过的
short-wavelength *n.* 短波长
fuel *v.* 加剧，推动
fossil fuels 化石燃料
substantially *adv.* 很大程度地
buildup *n.* 积聚，组合
trace gases 微量气体
methane *n.* 甲烷
nitrous oxide *n.* 一氧化二氮，笑气
chlorofluorocarbon *n.* 氯氟化碳

from the scientific community. One involves the role of **volcanic** activity and the other involves solar variability.

❿ The idea that explosive **volcanic eruptions** might alter the Earth's climate was first proposed many years ago. It is still regarded as a plausible explanation for some aspects of climatic variability. Explosive eruptions **emit** huge quantities of gases and **fine-grained debris** into the atmosphere. The greatest eruptions are sufficiently powerful to inject material high into the atmosphere, where it spreads around the globe and remains for many months or even years. Among the most persistent hypotheses of climate change have been those based on the idea that the Sun is a variable star and that its output of energy varies over time. The effect of such changes would seem direct and easily understood: Increases in solar output would cause the atmosphere to warm, and reductions would result in cooling. This notion can be used to explain climate change of any length or intensity. However, no major long-term variations in the total intensity of **solar radiation** have yet been measured outside the atmosphere. Such measurements were not even possible until satellite technology became available.

⓫ Carbon dioxide is a very significant component **meteorologically**. Carbon dioxide is influential because it is **transparent** to incoming **short-wavelength** solar radiation, but it is not transparent to some of the longer-wavelength outgoing the Earth radiation. A portion of the energy leaving the ground is absorbed by atmospheric CO_2. The Earth's tremendous industrialization of the past two centuries has been **fueled**—and still is fueled—by burning **fossil fuels**. The use of coal and other fuels is the most prominent means by which humans add CO_2 to the atmosphere, but it is not the only way. The clearing of forests also contributes **substantially**.

⓬ Carbon dioxide is not the only gas contributing to a global increase in temperature. In recent years atmospheric scientists have come to realize that the industrial and agricultural activities of people are causing a **buildup** of several **trace gases** that also play a significant role. The substances are called trace gases because their concentrations are so much smaller than that of carbon dioxide. The trace gases that are most important are **methane** (CH_4), **nitrous oxide** (N_2O), and **chlorofluorocarbons** (CFCs). These gases absorb wavelengths of outgoing radiation from the Earth that would otherwise escape into space. Although individually their impact is modest, taken together the effects of these trace gases play a significant role in warming the troposphere.

⓭ A significant impact of a **human-induced** global warming is a rise in sea level. Research indicates that sea level has risen between 10 and 23 centimeters (4 and 8 inches) over the past century and that this trend will continue at an accelerated rate. Some models indicate that the rise may approach or even exceed 50 centimeters (20 inches) by the end of the 21st century. If this happens, many beaches and wetlands will be eliminated, and coastal civilization would be severely **disrupted**. Climate models are in general agreement that one of the strongest signals of global warming should be a loss of **sea ice** in the Arctic. The sea-ice decline represents a combination of natural variability and human-induced global warming, with the latter becoming increasingly evident in coming decades. During the past decade, evidence has mounted to indicate that the extent of **permafrost** in the Northern Hemisphere has decreased, as would be expected under long-term warming conditions. As the permafrost thaws, lake water drains deeper into the ground, and lakes shrink or disappear. The human-induced increase in the amount of carbon dioxide in the atmosphere also has some serious implications for ocean chemistry and marine life. Recent studies show that about one third of the **human-generated** carbon dioxide currently ends up in the oceans. As a result, the ocean's pH[1] becomes lower, making seawater more **acidic**.

⓮ Climate in the 21st century, unlike the **preceding** thousand years, is not expected to be stable. Many of the changes will probably be gradual environmental shifts, **imperceptible** from year to year. Nevertheless, the effects, **accumulated** over decades, will have powerful economic, social, and political consequences. Despite our best efforts to understand future climate shifts, there is also the potential for "surprises". This simply means that, due to the complexity of the Earth's climate system, we might experience relatively sudden, unexpected changes or see some aspects of climate shift in an unexpected manner.

human-induced *adj.* 人为造成的,由人类导致的
disrupt *v.* 中断,扰乱
sea ice 海冰
permafrost *n.* 永久冻土
human-generated *adj.* 人产生的
acidic *adj.* 酸性的
preceding *adj.* 先前的,前面的
imperceptible *adj.* 难以察觉的,察觉不出的
accumulate *v.* 积累,积攒

Note

(1) **Pondus Hydrogenii** describes the degree of acidity and alkalinity of an aqueous solution, represented by the symbol pH. Under thermodynamic standard conditions, an aqueous solution with pH=7 is neutral, pH<7 is acidic, and pH>7 is alkaline.

Reading Comprehension

Keys

Directions Match the paragraphs with their appropriate subtitles.

1. Paragraphs 1 and 2 _____
2. Paragraph 3 _____
3. Paragraph 4 _____
4. Paragraphs 5 and 6 _____
5. Paragraphs 7 and 8 _____
6. Paragraphs 9 and 10 _____
7. Paragraphs 11 and 12 _____
8. Paragraphs 13 and 14 _____

A. The climate system
B. Some atmospheric basics
C. Natural causes of climate change
D. Some possible consequences of global warming
E. Climate and geology are linked
F. How is climate change detected?
G. Heating the atmosphere
H. Carbon dioxide, trace gases, and climate change

Passage B

Scan and read along the passage

Climate and Urban Development*

Scan and read along the vocabulary

rural *adj.* 农村的,乡村的
urban *adj.* 城市的,城镇的
heat island（城市上空气温偏高的）热岛
asphalt *n.* 沥青,柏油
evaporative cooling 蒸发冷却,蒸发降温

❶ For more than a hundred years, it has been known that cities are generally warmer than surrounding **rural** areas. This region of city warmth, known as the **urban heat island**[(1)], can influence the concentration of air pollution. However, before we look at its influence, let's see how the heat island actually forms.

❷ The urban heat island is due to industrial and urban development. In rural areas, a large part of the incoming solar energy is used in evaporating water from vegetation and soil. In cities, where less vegetation and exposed soil exist, the majority of the Sun's energy is absorbed by urban structures and **asphalt**. Hence, during warm daylight hours, less **evaporative cooling** in

---

* This passage is adapted from 新东方在线.托福阅读 TPO48-3:Climate and urban development[EB/OL]. [2022-01-20]. https://liuxue.koolearn.com/toefl/read/876-3478-q0.html.

cities allows surface temperatures to rise higher than in rural areas. The cause of the urban heat island is quite involved. Depending on the location, time of year, and time of day, any or all of the following differences between cities and their surroundings can be important: **albedo** (reflectivity of the surface), surface roughness, **emissions** of heat, emissions of moisture, and emissions of particles that affect net radiation and the growth of cloud **droplets**.

❸ At night, the solar energy (stored as vast quantities of heat in city buildings and roads) is slowly released into the city air. Additional city heat is given off at night (and during the day) by vehicles and factories, as well as by industrial and **domestic** heating and cooling units. The release of heat energy is retarded by the tall vertical city walls that do not allow **infrared radiation** to escape as readily as does the relatively level surface of the surrounding countryside. The slow release of heat tends to keep nighttime city temperatures higher than those of the faster-cooling rural areas. Overall, the heat island is strongest at night when **compensating** sunlight is absent; during the winter, when nights are longer and there is more heat generated in the city; and when the region is dominated by a high-pressure area with light winds, clear skies, and less humid air. Over time, increasing urban heat islands affect climatological temperature records, producing artificial warming in climatic records taken in cities. This warming, therefore, must be accounted for in interpreting climate change over the past century.

❹ The constant outpouring of pollutants into the environment may influence the climate of the city. Certain particles reflect solar radiation, thereby reducing the sunlight that reaches the surface. Some particles serve as **nuclei** upon which water and ice form. Water vapor **condenses** onto these particles when the relative humidity is as low as 70 percent, forming **haze** that greatly reduces visibility. Moreover, the added nuclei increase the frequency of city fog.

❺ Studies suggest that precipitation may be greater in cities than in the surrounding countryside; this phenomenon may be due in part to the increased roughness of city **terrain**, brought on by large structures that cause surface air to slow and gradually **converge**. This piling up of air over the city then slowly rises, much like toothpaste does when its tube is squeezed. At the same time, city

albedo *n.* 反照率,反射率
emissions *n.* 排放,排放物,辐射
droplets *n.* 小水滴
domestic *adj.* 家用的,家庭的
infrared radiation 红外辐射
compensating *adj.* 补偿的,平衡的
nuclei *n.* 核心,核子
condense *v.* 冷凝,凝结
haze *n.* 霾
terrain *n.* 地形,地势
converge *v.* (使)汇聚,集中

thermal *adj.* 热的，由热引起的，由温度变化引起的
breeze *n.* 微风
outskirt *n.* 郊区，市郊
dispersion *n.* 分散，散布，分布
inhibit *v.* 抑制，约束
downwind *adv.* 顺风，在下风

heat warms the surface air, making it more unstable, which enhances rising air motions, which, in turn, aids in forming clouds and thunderstorms. This process helps explain why both tend to be more frequent over cities.

❻ On clear still nights when the heat island is **pronounced**, a small **thermal** low-pressure area forms over the city. Sometimes a light **breeze**—called a country breeze—blows from the countryside into the city. If there are major industrial areas along the **outskirts**, pollutants are carried into the heart of town, where they tend to concentrate. Such an event is especially probable if vertical mixing and **dispersion** of pollutants are **inhibited**. Pollutants from urban areas may even affect the weather **downwind** from them.

Note

(1) **The Urban Heat Island** is a phenomenon that affects city dwellers worldwide. Heat island is an urban area in which significantly more heat is absorbed and retained than in surrounding areas. The higher temperatures experienced in urban areas compared to the surrounding countryside has enormous consequences for the health and wellbeing of people living in cities. The increased use of manmade materials and increased anthropogenic heat production are the main causes of the UHI. The UHI effect also leads to increased energy needs that further contribute to the heating of our urban landscape, and the associated environmental and public health consequences.

Reading Comprehension

 Keys

Directions Answer the following questions according to Passage B.

1. The word "involved" in Paragraph 2 is closest in meaning to   (   )
   A. uncertain.
   B. complicated.
   C. common.
   D. clear.

2. Paragraph 2 mentions all of the following as varying the importance of albedo and other factors EXCEPT   (   )
   A. seasons.
   B. soil depth.
   C. geographic location.

D. the time of day.

3. The word "retarded" in Paragraph 3 is closest in meaning to ( )
   A. disguised.
   B. added to.
   C. made possible.
   D. slowed down.

4. All of the following are mentioned in Paragraph 3 as contributing to an increase in the amount of heat within a city EXCEPT ( )
   A. home air conditioners.
   B. cars and trucks.
   C. street lights.
   D. factory buildings.

5. According to Paragraph 4, how do pollutants reduce the distance it is possible to see? ( )
   A. They increase the amount of sunlight that reaches the ground.
   B. They increase the relative humidity.
   C. They form particles that irritate the eye.
   D. They serve as nuclei around which water condenses.

6. Which of the sentences below best expresses the essential information in the highlighted sentence in Paragraph 5? ( )
   A. Until more studies are done, suggestions about the causes of precipitation in cities will focus on the roughness of terrain rather than on surface air and convergence.
   B. Certain phenomena of city landscapes, such as large structures, cause surface air to slow and converge, which brings a change in weather patterns to cities and rural areas.
   C. One reason why precipitation may be greater in cities than in the countryside is that large buildings that are found in cities cause surface air to slow and converge.
   D. Studies that focus on large structures, which are only partly responsible for the increased roughness of city terrain, are incomplete in their explanation of increased precipitation.

7. Why does the author mention "toothpaste" being squeezed from a tube in Paragraph 5?
   ( )
   A. To compare the movement of toothpaste from a tube to the movement of precipitation from clouds.
   B. To suggest that the process of cloud formation is a simple, everyday experience.
   C. To help the reader visualize the process of air movement over a city.
   D. To contrast the slow rising of air currents with the rapid squeezing of toothpaste.

8. The word "pronounced" in Paragraph 6 is closest in meaning to
   A. examined.
   B. relative.
   C. strongest.
   D. darkest.

9. The letters [A], [B], [C], and [D] indicate where the following sentence (in bold) could be added to the following part of the passage. Where would the sentence best fit? (   )

> **The resulting difference in atmosphere pressure between the city and the countryside can cause air to shift.**

On clear still nights when the heat island is pronounced, a small thermal low-pressure area forms over the city. [A] Sometimes a light breeze—called a country breeze—blows from the countryside into the city. [B] If there are major industrial areas along the outskirts, pollutants are carried into the heart of town, where they tend to concentrate. [C] Such an event is especially probable if vertical mixing and dispersion of pollutants are inhibited. [D] Pollutants from urban areas may even affect the weather downwind from them.

10. An introductory sentence (in bold) for a brief summary of the passage is provided below. Complete the summary by selecting the THREE answer choices that express the most important ideas in the passage. (   )

> **Cities are generally warmer than the surrounding countryside, a phenomenon known as the urban heat island.**

A. In the countryside, much solar energy is used in evaporation, but in the city this energy builds up as heat.

B. The urban heat island is strongest in the summer, when the days are long and the sunlight is intense.

C. Increased industrial and urban development has also increased average levels of humidity over the last century.

D. Heat and air are trapped in the irregular spaces between buildings, which creates the atmospheric conditions that result in storms and winds.

E. Pollution from cars and factories helps increase the amounts of fog and precipitation that occur in cities.

F. Country breezes blow pollutants put from the cities into the surrounding countryside.

Passage C

ENSO Events and Climate Change*

❶ Generated in the tropical Pacific, El Niño-Southern Oscillation (ENSO) events create a **far-reaching** system of climate **anomalies** that operate on a range of time scales important to society. ENSO influences extreme weather events such as drought, flooding, bushfires, and tropical **cyclone** activity across vast areas of the Earth, **adversely** affecting hundreds of millions of people in agriculturally important areas of Australasia, Africa, and the Americas. Despite being the dominant source of **inter-annual** climate variability, many characteristics of long-term changes in the frequency, duration, and magnitude of ENSO events remain unknown.

❷ ENSO is a **coupled** cycle in the atmosphere-oceanic system. It is an irregular phenomenon that alternates between its two phases, **El Niño**[1] and **La Niña**[2], approximately every 2–7 years. ENSO is a periodic reorganization of Sea Surface Temperature (SST) and atmospheric circulation in the tropical Pacific that results in vast redistributions of major rainfall-producing systems.

❸ Multi-century **paleoclimate** reconstructions derived from long proxy records, such as annually resolved tree-ring, coral, ice-core, and documentary records, provide a long-term context for assessing the significance of the apparently **anomalous** ENSO variability of the recent past. ENSO reconstruction is further complicated by the fact that proxies are not uniformly accurate in recording their local climate or **oceanographic** environment, and sometimes the narrow seasonal response of climate-sensitive proxies may not coincide with the **seasonality** of the local ENSO **teleconnection**.

❹ A number of proxy ENSO records were examined to isolate ENSO signals associated with both phases of the phenomenon. By using a variety of regional records, it is possible to capture more of

---

* This passage is adapted from GERGIS J L, FOWLER A M, 2009. A history of ENSO events since A.D. 1525: implications for future climate change[J]. Climatic Change, 92(3): 343-387.

spatial variability	空间变异
pre-instrumental	*adj.* 有仪器检测之前的
infancy	*n.* 初期,幼年,婴儿期
propagation	*n.* 传播,传送,蔓延
stationarity	*n.* 平稳性,稳定性
sensitivity	*n.* 敏感性
chronology	*n.* 年表,年代学
simulation	*n.* 模拟
indice	*n.* (古法语)指数,标记体
archaeologist	*n.* 考古学家
variance	*n.* 方差,变化幅度,差额
truncation	*n.* 截断
regression	*n.* (统计)回归
climate variability	气候变异

the **spatial variability** of ENSO more likely to be representative of large-scale ocean-atmosphere processes than is possible from single proxy analysis. Regional ENSO event signatures were compared, revealing large-scale trends in the frequency, magnitude, and duration of **pre-instrumental** ENSO. Clearly, multi-proxy ENSO reconstruction is still in its **infancy**. Abundant potential remains to characterize teleconnection patterns, low frequency changes, **propagation** signatures, and non-**stationarities** of large-scale ENSO behavior. It is well recognized that there is a need for high-quality proxies from key ENSO-affected regions, particularly from the western Pacific. It is essential that existing records from these regions be further developed and/or reviewed for their ENSO-**sensitivity** to allow regional dynamics of the western Pacific to be resolved.

❺ Using a variety of regional ENSO signals spanning both the east and west Pacific an extensive 478-year **chronology** of ENSO events was developed back to A.D. 1525. The large-scale (rather than regional) trends in the frequency, magnitude, and duration of pre-instrumental ENSO over the past five centuries were allowed. The most comprehensive La Niña event chronology complied to date is presented for the A.D. 1525–2002 period. This annual record of ENSO events can now be used as an independent means of verifying model **simulations**, continuous proxy reconstructions of ENSO **indices**, and a chronological control for **archaeologists** and social scientists studying human responses to past climate events.

❻ Significantly, none of the ENSO indices reconstructed to date are able to completely reproduce the **variance** exhibited by the instrumental record. This may reflect both the **truncation** of variance due to **regression**-type approaches to generating transfer functions as well as inherent limitations in the ability of paleoclimate proxies to fully resolve the magnitude of associated **climate variability**. This reinforces the fact that different ENSO reconstruction techniques have important biases and limitations to consider.

❼ Despite their limitations, reconstructions of past climate are unique in their ability to provide a long-term context for evaluating 20th century climate change. The unusual nature of late 20th century ENSO is evident. The height of La Niña activity occurred during the 16th and 19th centuries, while the 20th century is

identified as the peak period of El Niño activity. Overall, 55% of extreme El Niño event years reconstructed since A.D. 1525 occur within the 20th century. Although extreme ENSO events are seen throughout the 478-year ENSO reconstruction, approximately 43% of extreme and 28% of all **protracted** ENSO events reconstructed occur in the 20th century. Of particular note, the post-1940 period alone accounts for 30% of extreme ENSO years noted since A.D. 1525.

❽ This trend towards **enhanced** ENSO activity is noted in the IPCC's[3] fourth assessment report. Global mean temperatures are influenced by ENSO through the large exchanges of heat between the ocean and atmosphere. Observed changes in ENSO behavior may be more closely linked to global climate change than currently indicated by the IPCC's future climate **projections**. If this is the case, it is possible that extremes of the hydrological cycle that produce droughts, and floods during ENSO events may be enhanced under future global warming.

❾ It is important to note, however, that reconstructions of ENSO events alone are insufficient to clearly characterize past ENSO behavior. Reconstructions may give us a greater appreciation of the magnitude and nature of past ENSO behavior providing a much needed context for understanding recent observed changes required to **constrain** future climate change projections. Further research efforts should be directed into reconstructing ENSO variability across the entire frequency domain and modeling historic **synoptic** conditions. In this way, proxy reconstructions could be more readily used to constrain numerical experiments used to assess uncertainties in ENSO dynamics, like response of extreme events to natural/**anthropogenic** forcing.

❿ A recent comparison of a tropical SST reconstruction with two general circulation models indicates that the late 20th century is likely the warmest period in the tropics for the last 250 years and that this recent warming can only be explained by anthropogenic forcing. ENSO may operate differently in increasingly anthropogenically forced climate system. However, according to model ensemble results considered in the IPCC fourth assessment report, there is no **consistent** indication of how the **amplitude** or frequency of ENSO will operate under global

---

identify v. 查明，确认
protract v. 延长时间
enhance v. 增强，提高
projection n. 估算，预测
constrain v. 限制，约束
synoptic adj. 天气的
anthropogenic adj. 人为的
consistent adj. 一致的
amplitude n. 广度，阔度

availability n. 可用性,可得性
probability n. 概率,可能性
vulnerability n. 脆弱性
biophysical adj. 生物物理学的
water scarcity 水短缺,水荒
priority n. 优先事项,优先

warming. Model simulations of past ENSO variability are likely to be our best guide for projecting future patterns of temperature and rainfall caused by ENSO.

⑪ A greater understanding of ENSO is especially important for national assessments of the socio-economic risks associated with future climate change. As recognized in by the IPCC, the **availability** of observational data restricts the types of extremes events that can be analyzed for the identification of long-term changes. Extending the current time series of ENSO events beyond the 20th century may assist natural disaster managers to revise event **probabilities** used for assessing risk **vulnerability** of **biophysical** and socio-economic systems to threats like **water scarcity** or crop failures. Improvements in quantitative predictive tools for climate risk management, such as rainfall forecasts, commodity price projections and long-term trends in agricultural productivity are vital for providing early warnings of drought and crop failure risks. This is particularly relevant to policy makers and natural resource managers developing future climate change adaptation strategies. As evidence of stresses on water supply, agriculture, and natural ecosystems caused by climate change strengthens, studies into how ENSO will operate in a warmer climate should be a global research **priority**.

Notes

(1) **El Niño** is the warm phase of the El Niño–Southern Oscillation (ENSO) and is associated with a band of warm ocean water that develops in the central and east-central equatorial Pacific (approximately between the International Date Line and 120°W), including the area off the Pacific coast of South America.

(2) **La Niña** is the colder counterpart of El Niño, as part of the broader El Niño–Southern Oscillation (ENSO) climate pattern. During a La Niña period, the sea surface temperature across the eastern equatorial part of the central Pacific Ocean will be lower than normal by 3–5 ℃(5.4 – 9 F).

(3) **The Intergovernmental Panel on Climate Change** (IPCC) is an intergovernmental body of the United Nations responsible for advancing knowledge on human-induced climate change. It was established in 1988 by the World Meteorological Organization (WMO) and the United Nations Environment Programme (UNEP), and later endorsed by United Nations General Assembly. Headquartered in Geneva, Switzerland, it is composed of 195 member states.

Reading Comprehension

 Keys

Directions Answer the following questions according to Passage C.

1. According to Paragraph 1, what are the adverse effects of ENSO?

2. What are the two phases of ENSO?

3. What makes ENSO reconstruction further complicated?

4. What is the significance of the annual record of ENSO events?

5. What are the limitations of the ENSO indices reconstructed to date?

6. Under what conditions are extremes of the hydrological cycle during ENSO events may be enhanced?

7. According to Paragraph 9, where should further research efforts be directed?

8. How does a greater understanding of ENSO help nations assess the socio-economic risks associated with future climate change?

Exercises

Vocabulary

I. Technical words

Directions Match the technical terms with their definitions.

A. proxy
B. paleoclimatology
C. troposphere
D. permafrost
E. magnitude
F. teleconnection
G. chronology
H. hydrology

a. land that is permanently frozen to a great depth
b. something that you use to represent something else that you are trying to measure or calculate
c. the study of climates of the geological past
d. the scientific study of the Earth's water, especially its movement in relation to land
e. the lowest layer of the Earth's atmosphere, between the surface of the Earth and about 6-10 kilometers above the surface
f. the great size or importance of something; the degree to which something is large or important
g. climate anomalies being related to each other at large distances (typically thousands of kilometers)
h. an account or record of the times and the order in which a series of past events took place

II. Academic vocabulary

Directions Fill in the blanks with appropriate forms of the verbs given below. Each word can be used only once.

Verb	Definition in Academic Writing
assess	to make a judgement about the nature or quality of sb./sth.
analyze	to examine the nature or structure of sth., especially by separating it into its parts, in order to understand or explain it
constrain	to restrict or limit sb./sth.
enhance	to increase or further improve the good quality, value or status of sb./sth.

explain	to give details about sth. or describe it so that it can be understood; to give people reasons for sth. that has happened
identify	to recognize sb./sth. and be able to say who or what they are
note	to mention sth. because it is important or interesting
operate	to work in a particular way
recognize	to admit or to be aware that sth. exists or is true
review	to carefully examine or consider sth. again, especially so that you can decide if it is necessary to make changes

1. In a recent article, the impact of climate change on the process of creating a pluralistic society was _____.
2. A new method is developed to _____ S-wave velocity structures of the shallow crust based on frequency-dependent amplitudes of direct P-waves in P-wave receiver functions (P–RFs).
3. Theories that petroleum is not formed by the transformation of organic matter in sediments have already been _____ and are examined in more detail.
4. The aim of this paper is to _____ the importance of Water Demand Management (WDM) strategy to the improvement of water supply and sanitation in Nigeria.
5. *Chemical Geology* publishes geochemical studies of fundamental and broad significance to _____ the understanding of processes of the Earth and the solar system.
6. The paper proposes a framework for setting the institutional groundwork for _____ climate migrants, focusing on the most vulnerable, promoting targeted research and policy agendas, and situating policies within a comprehensive strategy.
7. We _____ the implications of these links for adaptation finance and what the literature tells us about the balance between adaptation and mitigation.
8. Natural variation in the relative abundances of elements and isotopes of various elements can be used as a tool to _____ the mechanisms behind geological systems on the Earth and in the universe.
9. A mass-independent oxygen isotopic composition has been suggested to be a useful tool to _____ sulfate aerosol formation pathways.
10. Geology provides clear evidence of the global flood of Noah, which strongly supports a young earth and goes against the long ages the theory of evolution requires to _____.

Sentence Structure

Understand adverbial clauses in English sentences.

An adverbial clause is a dependent clause that functions as an adverb within a sentence.

Unlike adverbs, adverbial clauses modify whole clauses rather than just a verb. An adverbial clause begins with a subordinating conjunction such as *when, where, because*, or *although*. Adverbial clauses are traditionally classified according to their meaning. In academic texts, adverbial clauses of time, place, reason, concession, condition, contrast, and purpose are often used.

> For example,
> The industrial revolution only really got going *when* coal began to be used widely. (time)
> Soil forms *where* the geosphere, the atmosphere, the hydrosphere, and the biosphere meet. (place)
> Ideally, disposal will not require transportation over long distances *because* the transport of hazardous materials is risky. (reason)
> *Although* color is an obvious feature of a mineral, it is often an unreliable diagnostic property. (concession)
> *If* weathering has been going on for a comparatively short time, the parent material strongly influences the characteristics of the soil. (condition)
> The portion of the soil consisting of pore spaces that allow for the circulation of air and water is as vital *as* the solid soil constituents. (contrast)
> It must be engineered so *that* it cannot be easily entered, damaged, or sabotaged. (purpose)

Directions Analyze the following sentences by: 1) underlining the subordinating conjunctions that introduce the adverbial clause(s), 2) determining what adverbial clauses are included.

1. The atmosphere rapidly thins as you travel away from the Earth, until there are too few gas molecules to detect. (Paragraph 6, Passage A)

2. The substances are called trace gases because their concentrations are so much smaller than that of carbon dioxide. (Paragraph 12, Passage A)

3. The release of heat energy is retarded by the tall vertical city walls that do not allow infrared radiation to escape as readily as does the relatively level surface of the surrounding countryside. (Paragraph 3, Passage B)

4. Although extreme ENSO events are seen throughout the 478-year ENSO reconstruction, approximately 43% of extreme and 28% of all protracted ENSO events reconstructed occur in the 20th century. (Paragraph 7, Passage C)

Unit 3　Atmospheric and Climate Sciences

Translation

Ⅰ. **Translate the following English sentences into Chinese. Pay attention to the adverbial clause in each sentence.**

1. Although these gases are the most plentiful components of air and are of great significance to life on the Earth, they are of little or no importance in affecting weather phenomena.

2. If this happens, many beaches and wetlands will be eliminated and coastal civilization would be severely disrupted.

3. Water vapor condenses onto these particles when the relative humidity is as low as 70 percent, forming haze that greatly reduces visibility.

4. As evidence of stresses on water supply, agriculture and natural ecosystems caused by climate change strengthens, studies into how ENSO will operate in a warmer climate should be a global research priority.

Ⅱ. **Translate the following Chinese paragraph into English.**

显然,气候变化问题越来越严重,这是被世界上许多科学家和研究人员以及众多政府承认的。最近的数据显示,世界海洋有记录以来,2019年是最热的。更令人恐慌的消息是,2020年是有记录以来第二热的一年,而陆地地区是最热的。2020年陆地和海洋表面平均温度比1900年至2000年期间的平均温度高0.98 ℃。此外,有记录以来最热的10年出现在2005年之后。

Unit 4　Soil Science

Soil covers most of the land surface. Along with air and water, it is one of our most indispensable resources. Also, like air and water, soil is taken for granted by many of us. Soil has accurately been called the bridge between life and the inanimate world. All life—the entire biosphere owes its existence to a dozen or so elements that must ultimately come from the Earth's crust. Once weathering and other processes create soil, plants carry out the intermediary role of assimilating the necessary elements and making them available to animals, including humans. In this unit, you will read about: 1) the most important factors that control soil formation, 2) certain minerals required by plants for normal growth and development, and 3) soil health, the continued capacity of soil to function as a vital living ecosystem.

Passage A

Scan and read along the passage

Scan and read along the vocabulary

regolith *n.* 风化层, 土被

A Brief Introduction to Soil*

❶　When the Earth is viewed as a system, soil is referred to as an interface—a common boundary where different parts of a system interact. This is an appropriate designation because soil forms where the geosphere, the atmosphere, the hydrosphere, and the biosphere meet.

What is soil?

❷　With few exceptions, the Earth's land surface is covered by **regolith**, the layer of rock and mineral fragments produced by weathering. Some would call this material soil, but soil is more than an accumulation of weathered debris. Soil is a combination of

---

* The passage is adapted from LUTGENS F K, TARBUCK E J, TASA D, 2011. Essentials of geology[M]. 11th ed. Upper Saddle River, N J: Pearson Prentice Hall.

mineral and organic matter, water, and air—that portion of the regolith that supports the growth of plants. Although the proportions of the major components in soil vary, the same four components are always present to some extent. About one-half of the total volume of a good quality surface soil is a mixture of **disintegrated** and **decomposed** rock (mineral matter) and **humus**, the decayed remains of animal and plant life (organic matter). The remaining half consists of **pore** spaces among the solid particles where air and water circulate.

❸ Although the mineral portion of the soil is usually much greater than the organic portion, humus is an essential component. In addition to being an important source of plant nutrients, humus enhances the soil's ability to retain water. Because plants require air and water to live and grow, the portion of the soil consisting of pore spaces that allow for the circulation of these fluids is as vital as the solid soil constituents. Soil water is far from "pure" water; instead, it is a complex solution containing many **soluble** nutrients. Soil water not only provides the necessary moisture for the chemical reactions that sustain life, it also supplies plants with nutrients in a form they can use. The pore spaces that are not filled with water contain air. This air is the source of necessary oxygen and carbon dioxide for most **microorganisms** and plants that live in the soil.

Controls of Soil Formation

❹ Soil is the product of the complex **interplay** of several factors. The most important of these are parent material[1], time, climate, plants and animals, and topography. Although all of these factors are interdependent, their roles will be examined separately.

❺ The source of the weathered mineral matter from which soils develop is called the parent material and is a major factor influencing a newly forming soil. Gradually it undergoes physical and chemical changes as the processes of soil formation progress. Parent material might be the underlying **bedrock**, or it can be a layer of unconsolidated deposits, as in a stream valley. When the parent material is bedrock, the soils are termed **residual** soils. By contrast, those developed on unconsolidated sediment are called transported soils. Note that transported soils form in place on parent materials that have been carried from elsewhere and deposited by gravity, water, wind, or ice.

---

disintegrated *adj.* 破碎的
decomposed *adj.* 分解的
humus *n.* 腐殖质,腐殖土
pore *n.* (皮肤上的)毛孔,(植物的)气孔
soluble *adj.* 可溶的,可解决的
microorganism *n.* 微生物,微小动植物
interplay *n.* 相互影响,相互作用
bedrock *n.* 基岩,基本原理
residual *adj.* 残留的,(土壤)残余的

granite n. 花岗岩
marble n. 大理石
vegetation n. (总称)植物,植被
dissimilar adj. 不同的
reinforce v. 加强,加固
predominate v. (数量上)占优势,占支配地位

❻ The nature of the parent material influences soils in two ways. First, the type of parent material affects the rate of weathering and thus the rate of soil formation. (Consider the weathering rates of **granite** versus **marble**.) Also, because unconsolidated deposits are already partly weathered and provide more surface area for chemical weathering, soil development on such material usually progresses more rapidly. Second, the chemical make-up of the parent material affects the soil's fertility. This influences the character of the natural **vegetation** the soil can support.

❼ At one time the parent material was thought to be the primary factor causing differences among soils. However, soil scientists came to understand that other factors, especially climate, are more important. In fact, it was found that similar soils often develop from different parent materials and that **dissimilar** soils develop from the same parent material. Such discoveries **reinforce** the importance of the other soilforming factors.

❽ Time is an important component of every geological process, and soil formation is no exception. The nature of soil is strongly influenced by the length of time that processes have been operating. If weathering has been going on for a comparatively short time, the parent material strongly influences the characteristics of the soil. As weathering processes continue, the influence of parent material on soil is overshadowed by the other soil-forming factors, especially climate. The amount of time required for various soils to evolve cannot be specified, because the soil-forming processes act at varying rates under different circumstances. However, as a rule, the longer a soil has been forming, the thicker it becomes, and the less it resembles the parent material.

❾ Climate is the most influential control of soil formation. Just as temperature and precipitation are the climatic elements that influence people the most, so too are they the elements that exert the strongest impact on soil formation. Variations in temperature and precipitation determine whether chemical or mechanical weathering **predominates**. They also greatly influence the rate and depth of weathering. For instance, a hot, wet climate may produce a thick layer of chemically weathered soil in the same amount of time that a cold, dry climate produces a thin mantle of mechanically weathered debris. Also, the amount of precipitation influences the degree to which various materials are removed from the soil, thereby affecting soil fertility. Finally, climatic conditions are important factors controlling the type of plant and animal life present.

❿ The biosphere plays a vital role in soil formation. The types and abundance of organisms present have a strong influence on the physical and chemical properties of a soil. In fact, for well-developed soils in many regions, the significance of natural vegetation in influencing soil type is frequently implied in the description used by soil scientists. Such phrases as *prairie soil*, *forest soil*, and *tundra soil* are common. Plants and animals furnish organic matter to the soil. Certain bog soils are composed almost entirely of organic matter, whereas desert soils may contain only a tiny percentage. Although the quantity of organic matter varies substantially among soils, it is the rare soil that completely lacks it.

⓫ The primary source of organic matter is plants, although animals and the uncountable microorganisms also contribute. When organic matter decomposes, nutrients are supplied to plants, as well as to animals and microorganisms living in the soil. Consequently, soil fertility depends in part on the amount of organic matter present. Furthermore, the decay of plant and animal remains causes the formation of various organic acids. These complex acids hasten the weathering process. Organic matter also has a high water-holding ability and thus aids water **retention** in a soil. Microorganisms, including **fungi**, bacteria, and single-celled **protozoa**, play an active role in the decay of plant and animal remains. The end product is humus, a material that no longer resembles the plants and animals from which it was formed. In addition, certain microorganisms aid soil fertility because they have the ability to convert atmospheric nitrogen into soil nitrogen.

⓬ Earthworms and other **burrowing** animals act to mix the mineral and organic portions of a soil. Earthworms, for example, feed on organic matter and thoroughly mix soils in which they live, often moving and enriching many tons per acre each year. Burrows and holes also aid the passage of water and air through the soil.

prairie n.（北美的）大草原
tundra n. 苔原，土地带
retention n. 保持，保留
fungi n. 真菌，菌类，蘑菇（fungus 的复数）
protozoa n. 原生动物
burrow v. 掘地洞，挖地道

nonexistent *adj.* 不存在的
waterlog *v.* 使(船等)进水,使浸透水
saturate *v.* 浸透,使饱和
retard *v.* 减慢,受到阻滞
optimum *adj.* 最优的,最适宜的
undulating *adj.* 波状的,波浪起伏的
infiltration *n.* 渗透,渗透物

⓭ The lay of the land can vary greatly over short distances. Such variations in topography can lead to the development of a variety of localized soil types. Many of the differences exist because the length and steepness of slopes have a significant impact on the amount of erosion and the water content of soil.

⓮ On steep slopes, soils are often poorly developed. In such situations little water can soak in; as a result, soil moisture may be insufficient for vigorous plant growth. Further, because of accelerated erosion on steep slopes, the soils are thin or **nonexistent**.

⓯ In contrast, **waterlogged** soils in poorly drained bottomlands have a much different character. Such soils are usually thick and dark. The dark color results from the large quantity of organic matter that accumulates because **saturated** conditions **retard** the decay of vegetation. The **optimum** terrain for soil development is a flat-to-**undulating** upland surface. Here we find good drainage, minimum erosion, and sufficient **infiltration** of water into the soil. Slope orientation, or the direction a slope is facing, also is significant. In the midlatitudes of the northern Hemisphere, a south-facing slope receives a great deal more sunlight than a north-facing slope. In fact, a steep north-facing slope may receive no direct sunlight at all. The difference in the amount of solar radiation received causes substantial differences in soil temperature and moisture, which in turn influence the nature of the vegetation and the character of the soil.

⓰ Although we have dealt separately with each of the soil-forming factors, remember that all of them work together to form soil. No single factor is responsible for a soil's character. Rather, it is the combined influence of parent material, time, climate, plants and animals, and topography that determines this character.

⓱ *Soil*—that portion of the *regolith* (the layer of rock and mineral fragments produced by weathering) that supports the growth of plants—is a combination of mineral and organic matter, water, and air. About half of the total volume of a good-quality soil is a mixture of disintegrated and decomposed rock (mineral matter), and *humus* (the decayed remains of animal and plant life); the remaining half consists of pore spaces, where air and water circulate. The most important factors that control soil formation are *parent material*, *time*, *climate*, *plants and animals*, and *topography*. Soil-forming processes operate from the surface downward and produce zones or layers in the soil called *horizons*. From the surface downward, the soil horizons are respectively designated as O (largely organic matter), A (largely mineral matter), E (where the fine soil components and soluble materials have been removed by eluviation and leaching), B (*subsoil*, often referred to as the *zone of accumulation*), and C (partially altered parent material). Together the O and A horizons make up what is commonly called the *topsoil*[2].

Notes

(1) **The parent material** for residual soils is the underlying bedrock, whereas transported soils form on unconsolidated deposits. Also note that as slopes become steeper, soil becomes thinner.

(2) **Topsoil** is the extreme upper part of the Earth's surface, extending downward only 5.08–30.48 centimeters. It usually takes between 80 and 400 years for soil-forming processes to create 1 cm of topsoil. Topsoil is inextricably intertwined with ecosystem stability, because it contains the necessary minerals and nutrients that living things—including the plants that directly or indirectly support thousands of species—require. Formed through natural processes, it has multiple uses and varies in terms of composition. People often add different items, such as manure, to make it more fertile and suitable for specific needs. Environmentalists have concerns about how to sustain it and keep it free from contamination.

Reading Comprehension

  Keys

Directions Answer the following questions according to Passage A.

1. How is regolith different from soil?

2. List the five basic controls of soil formation.

3. Which factor is the most influential in soil formation?

4. How might the direction a slope is facing influence soil formation?

5. What is the parent material?

6. In what ways does the nature of the parent material influence soil formation?

7. How does the length of time affect soils in general?

8. How does the parent material influence soils?

Passage B

Scan and read along the passage

Scan and read along the vocabulary

nitrate *n.* 硝酸盐

Minerals and Plants*

❶ Research has shown that certain minerals are required by plants for normal growth and development. The soil is the source of these minerals, which are absorbed by the plant with the water from the soil. Even nitrogen, which is a gas in its elemental state, is normally absorbed from the soil as **nitrate** ions. Some soils are notoriously deficient in micro-nutrients and are therefore unable to support

---

* This passage is adapted from 坦途网托福考试频道, 2018.托福阅读专项练习 TPO05: Minerals and plants[EB/OL]. [2022-04-22]. https://mobile.tantuw.com/toefl/detail-67906.

most plant life. So-called **serpentine** soils, for example, are deficient in calcium, and only plants able to tolerate low levels of this mineral can survive. In modern agriculture, mineral depletion of soils is a major concern, since harvesting crops interrupts the recycling of nutrients back to the soil.

❷   Mineral deficiencies can often be detected by specific symptoms such as **chlorosis** (loss of **chlorophyll** resulting in yellow or white leaf tissue), **necrosis** (isolated dead patches), **anthocyanin** formation (development of deep red **pigmentation** of leaves or stem), stunted growth, and development of woody tissue in an **herbaceous** plant. Soils are most commonly deficient in nitrogen and **phosphorus**. Nitrogen-deficient plants exhibit many of the symptoms just described. Leaves develop chlorosis; stems are short and slender; and anthocyanin discoloration occurs on stems, petioles, and lower leaf surfaces. **Phosphorus**-deficient plants are often stunted, with leaves turning a characteristic dark green, often with the accumulation of anthocyanin. Typically, older leaves are affected first as the phosphorus is mobilized to young growing tissue. Iron deficiency is characterized by chlorosis between veins in young leaves.

❸   Much of the research on nutrient deficiencies is based on growing plants hydroponically, that is, in soilless liquid nutrient solutions. This technique allows researchers to create solutions that selectively omit certain nutrients and then observe the resulting effects on the plants. **Hydroponics** has applications beyond basic research, since it facilitates the growing of greenhouse vegetables during winter. **Aeroponics**, a technique in which plants are suspended and the roots misted with a nutrient solution, is another method for growing plants without soil.

❹   While mineral deficiencies can limit the growth of plants, an overabundance of certain minerals can be toxic and can also limit growth. **Saline** soils, which have high concentrations of sodium **chloride** and other salts, limit plant growth, and research continues to focus on developing salt-tolerant varieties of agricultural crops. Research has focused on the toxic effects of heavy metals such as lead, **cadmium, mercury,** and **aluminum**; however, even copper and zinc, which are essential elements, can become toxic in high concentrations. Although most plants cannot survive in these soils certain plants have the ability to tolerate high

serpentine n. 蛇纹石,蛇形之物
    adj. 蜿蜒的,阴险的
chlorosis n. 变色病,萎黄病
chlorophyll n. 叶绿素
necrosis n. 坏疽,骨疽
anthocyanin n. 花色素苷
pigmentation n. 色素淀积,染色
herbaceous adj. 草本的,叶状的
phosphorus n. 磷
hydroponics n. 水培,无土栽培
aeroponics n. 空气种植法
saline adj. 盐的,含盐的,含镁盐类的
chloride n. 氯化物
cadmium n. (化学元素)镉
mercury n. 汞,水银
aluminum n. 铝

cobalt *n.* 钴，钴类颜料

manganese *n.* （化学元素）锰

mustard *n.* 芥末酱，芥末黄

spurge *n.* 大戟，大戟树

legume *n.* 豆类，豆科植物

microbial *adj.* 微生物的，由细菌引起的

pathogen *n.* 病原体，致病菌

phytoremediation *n.* 植物修复

compost *n.* 堆肥，混合物；*v.* 堆肥，施堆肥

excavation *n.*（对古物的）发掘，挖掘

alpine *adj.* 阿尔卑斯山的，高山的

pennycress *n.* 菥蓂

selenium *n.* 硒

levels of these minerals.

❺ Scientists have known for some time that certain plants, called hyperaccumulators, can concentrate minerals at levels a hundredfold or greater than normal. A survey of known hyperaccumulators identified that 75 percent of them amassed nickel, **cobalt**, copper, zinc, **manganese**, lead, and cadmium are other minerals of choice. Hyperaccumulators run the entire range of the plant world. They may be herbs, shrubs, or trees. Many members of the **mustard** family, **spurge** family, **legume** family, and grass family are top hyperaccumulators. Many are found in tropical and subtropical areas of the world, where accumulation of high concentrations of metals may afford some protection against plant-eating insects and **microbial pathogens**.

❻ Only recently have investigators considered using these plants to clean up soil and waste sites that have been contaminated by toxic levels of heavy metals—an environmentally friendly approach known as **phytoremediation**. This scenario begins with the planting of hyperaccumulating species in the target area, such as an abandoned mine or an irrigation pond contaminated by runoff. Toxic minerals would first be absorbed by roots but later relocated to the stem and leaves. A harvest of the shoots would remove the toxic compounds off site to be burned or **composted** to recover the metal for industrial uses. After several years of cultivation and harvest, the site would be restored at a cost much lower than the price of **excavation** and reburial, the standard practice for remediation of contaminated soils. For examples, in field trials, the plant **alpine pennycress** removed zinc and cadmium from soils near a zinc smelter, and Indian mustard, native to Pakistan and India, has been effective in reducing levels of **selenium** salts by 50 percent in contaminated soils.

Reading Comprehension

Keys

Directions Answer the following questions according to Passage B.

1. According to Paragraph 1, what is true of plants that can grow in serpentine soil? (   )

    A. They absorb micronutrients unusually well.

    B. They require far less calcium than most plants do.

    C. They are able to absorb nitrogen in its elemental state.

D. They are typically crops raised for food.
2. According to Paragraph 2, which of the following symptoms occurs in phosphorus-deficient plants but not in plants deficient in nitrogen or iron? ( )
   A. Chlorosis on leaves.
   B. Change in leaf pigmentation to a dark shade of green.
   C. Short, stunted appearance of stems.
   D. Reddish pigmentation on the leaves or stem.
3. According to Paragraph 2, a symptom of iron deficiency is the presence in young leaves of ( )
   A. deep red discoloration between the veins.
   B. white or yellow tissue between the veins.
   C. dead spots between the veins.
   D. characteristic dark green veins.
4. The word "facilitates" in Paragraph 3 is closest in meaning to ( )
   A. slows down.
   B. affects.
   C. makes easier.
   D. focuses on.
5. According to Paragraph 3, what is the advantage of hydroponics for research on nutrient deficiencies in plants? ( )
   A. It allows researchers to control what nutrients a plant receives.
   B. It allows researchers to observe the growth of a large number of plants simultaneously.
   C. It is possible to directly observe the roots of plants.
   D. It is unnecessary to keep misting plants with nutrient solutions.
6. Why does the author mention "herbs" "shrubs" and "trees" in Paragraph 5? ( )
   A. To provide examples of plant types that cannot tolerate high levels of harmful minerals.
   B. To show why so many plants are hyperaccumulators.
   C. To help explain why hyperaccumulators can be found in so many different places.
   D. To emphasize that hyperaccumulators occur in a wide range of plant types.
7. Which of the sentences below best expresses the essential information in the highlighted sentence in Paragraph 6? ( )
   A. Before considering phytoremediation, hyperaccumulating species of plants local to the target area must be identified.
   B. The investigation begins with an evaluation of toxic sites in the target area to determine the extent of contamination.
   C. The first step in phytoremediation is the planting of hyperaccumulating plants in the area to be cleaned up.
   D. Mines and irrigation ponds can be kept from becoming contaminated by planting hyperaccumulating species in targeted areas.

8. It can be inferred from Paragraph 6 that compared with standard practices for remediation of contaminated soils, phytoremediation ( )

   A. does not allow for the use of the removed minerals for industrial purposes.

   B. can be faster to implement.

   C. is equally friendly to the environment.

   D. is less suitable for soils that need to be used within a short period of time.

9. The letters [A], [B], [C], and [D] indicate where the following sentence (in bold) could be added to the following part of the passage. Where would the sentence best fit? ( )

   Certain minerals are more likely to be accumulated in large quantities than others.

   [A] A survey of known hyperaccumulators identified that 75 percent of them amassed nickel, cobalt, copper, zinc, manganese, lead, and cadmium are other minerals of choice. [B] Hyperaccumulators run the entire range of the plant world. [C] They may be herbs, shrubs, or trees. [D] Many members of the mustard family, spurge family, legume family, and grass family are top hyperaccumulators.

10. An introductory sentence (in bold) for a brief summary of the passage is provided below. Complete the summary by selecting the THREE answer choices that express the most important ideas in the passage. ( )

    Plants need to absorb certain minerals from the soil in adequate quantities for normal growth and develop ment.

    A. Some plants are able to accumulate extremely high levels of certain minerals and thus can be used to clean up soils contaminated with toxic levels of these minerals.

    B. Though beneficial in lower levels, high levels of salts, other minerals, and heavy metals can be harmful to plants.

    C. When plants do not absorb sufficient amounts of essential minerals, characteristic abnormalities result.

    D. Because high concentrations of sodium chloride and other salts limit growth in most plants, much research has been done in an effort to develop salt-tolerant agricultural crops.

    E. Some plants can tolerate comparatively low levels of certain minerals, but such plants are of little use for recycling nutrients back into depleted soils.

    F. Mineral deficiencies in many plants can be cured by misting their roots with a nutrient solution or by transferring the plants to a soilless nutrient solution.

Passage C

 *Scan and read along the passage*

What is Soil Health*

 Scan and read along the vocabulary

❶ Soil health is the continued capacity of soil to function as a vital living ecosystem that sustains plants, animals, and humans, and connects agricultural and soil science to policy, **stakeholder** needs, and sustainable supply-chain management. Historically, soil assessments focused on crop production, but, today, soil health also includes the role of soil in water quality, climate change, and human health. However, quantifying soil health is still dominated by chemical indicators, despite growing appreciation of the importance of soil biodiversity, owing to limited functional knowledge and lack of effective methods.

❷ Soil is a complex system at the **intersection** of the atmosphere, lithosphere, hydrosphere, and biosphere that is critical to food production and key to sustainability through its support of important societal and eco-system services. It is in this context that the concept of soil health emerged in the early 2000s and, today, has linkages to the emerging "One Health" concept, in which the health of humans, animals, and the environment are all connected.

❸ The terminology, concept, and operationalization of soil health are still evolving. It is now defined by most agencies, such as the US Department of Agriculture, as "the continued capacity of soil to function as a vital living ecosystem that sustains plants, animals, and humans". Several other related concepts exist, including soil fertility, soil quality, and soil security, which also emphasize the role or functioning of soil in society, ecosystems, and/or agriculture. The narrowest of these terms is soil fertility, which refers to the role of soil in crop production. Soil fertility is managed by farmers at the field scale for the purpose of **cost-effective** crop production and entirely focuses on growing food, fuel, and fiber for human use.

stakeholder *n.* 股东,利益相关者
intersection *n.* 交接(点或线),相交,交汇点(尤指道路)
cost-effective *adj.* 有成本效益的,划算的

---

* The passage is adapted from LEHMANN J, BOSSIO D A, et al., 2020. The concept and future prospects of soil health [J]. Nature Reviews Earth & Environment, 1(10): 544–553.

encompass v. 包含，包括
align v. 公开支持，与……结盟，使平行，加入
temporal adj. 时间的，与时间有关的

❹ Soil quality is the historic origin of the term soil health and describes the ability of a soil to function for agriculture and its immediate environmental context. Soil quality, therefore, includes soil effects on water quality, plant, and animal health within entire eco-systems. Although the terms are often used synonymously, we argue that soil health is distinct from soil quality, as the scope of soil health extends beyond human health to broader sustainability goals that include planetary health, whereas the scope of soil quality usually focuses on ecosystem services with reference to humans.

❺ Soil security, introduced in 2012, is the most recent and broadest term of the four and **encompasses** soil health, using the term "soil condition" to describe the manageable properties of soil. Soil security relates to the need for access to soil ecosystem services to be on the same level as other human rights and is, therefore, often used in a policy context, encompassing human culture, capital, and legal aspects of soil management. Importantly, soil security allows for productive conversation about soil as a common good, similar to water and air, rather than only as private property (as in soil fertility and quality). We believe that this view must be moved to the center of the debate about the role of soils in sustainability and governance.

❻ Soil health encompasses scales, stakeholders, functions, and assessment tools relevant to soil quality and fertility, and shares some of the policy dimension of soil security, going beyond a focus on only crop production or other explicitly human benefits. The multidimensionality of the soil-health concept allows for soil-management goals to be **aligned** with sustainability goals, and should provide the foundation to consider a large number of stakeholders, functions, and spatial and **temporal** scales. One of the most important achievements of the soil-health framework (initially under the term soil quality) is the addition of an urgently needed biological perspective to soil management in order to address longer-term sustainability challenges for crop production. A biological perspective is also critical to expanding soil assessment and management to address concerns over biodiversity, water quality, climate, recreation, and human and planetary health beyond humans.

❼ The historical uneasiness with which scientists have embraced the concept of soil health is due to the challenges of defining soil health in a way that allows for a universal quantitative assessment that encompasses all of its ecosystem services, including human health. Reasons for this challenge include soil **heterogeneity**, the site-specific nature of soil management and the varying ecosystem services that have sometimes conflicting or competing needs. Nevertheless, there has been widespread interest amongst researchers, policymakers, and stakeholders in the use of the soil-health concept.

❽ In addition to managing physico-chemical soil properties for plant production, soil health considers the interactions between plants and soil **microbial** communities around roots, which can promote or reduce plant growth. Promoting a soil microbiome for high plant production requires management of microbial abundance and activity, community composition and specific functions. For example, organic amendments (such as compost) can foster increased **resilience** to plant pathogens through promotion of beneficial microorganisms. In many cases, higher organic matter content through higher amendments or reduced **tillage** increases biodiversity, which is expected to improve crop resilience. However, there are exceptions to these trends—reducing tillage, for example, can reduce crop yields in some instances, with follow-on reductions of soil organic carbon.

❾ Managing soil health to promote good water quality includes retaining pollutants and others in the soil, **buffering** against them and **biotically** transforming them. Increasing soil organic matter will retain heavy metals and organic **toxins**, some of which show nearly irreversible adsorption to organic matter. Using buffer zones, such as vegetative filter strips near agricultural areas or constructed wetlands, can slow the migration of nitrate, **phosphate** or **pesticide** contamination to water. Soil biota can transform organic pollutants, such as the common hydrocarbon **toluene**, into harmless compounds. Therefore, both organic-matter content and microbial activity, key properties of soil health, improve the quality of the water that is draining soil.

❿ Soil health of urban soils has not yet received sufficient recognition but can contain an even wider range of contaminants than agricultural soils, and many urban soils have also been

heterogeneity n. 异质性，非均匀性
microbial adj. 微生物的，由细菌引起的
resilience n. 恢复力，(橡胶等的)弹性
tillage n. 耕作，耕种
buffer n. 缓存，缓冲器
 v. 缓解，存储
biotical adj. 关于生命的，生物的
toxin n. 毒素，毒质
phosphate n. 磷酸盐
pesticide n. 杀虫剂，农药
toluene n. 甲苯

arsenic *n.* 砷,砒霜
abiotic *adj.* 非生物的,无生命的
mycotoxin *n.* 霉菌毒素,真菌毒素
parasitic *adj.* 寄生的
helminthiasis *n.* 蠕虫病,肠虫病
gastrointestinal *adj.* 胃肠的
municipality *n.* 市政当局,自治市
versatility *n.* 多功能性,多才多艺

modified to an extent that water can drain either very quickly or not at all. Soil-health management in urban soils must, therefore, balance eliminating surface run-off against retaining water and pollutants by reduced drainage. A combination of managing physical retention with biological transformation of pollutants through high soil biodiversity is the goal of bioretention and constructed soils to provide clean drinking water.

⓫ Human health depends, to a great extent, on soil health, including and going beyond the obvious connection between soil and human health through crop production. Similarly important is the type of crop and its nutritional content; soils with greater micronutrient availability are related to lower malnutrition and higher soil organic matter improves the nutritional value of crops. In addition to these relatively well-known properties, the nutritional value of crops can also depend on robust soil biodiversity, which can enhance micronutrient bioavailability to crops and suppress soil-borne plant disease, as well as affecting taste, food storage, and food preparation.

⓬ Soils can also negatively impact human health. For example, soil pollutants can contaminate produce through direct contact or dust, suspension or rainsplash. Some compounds, such as **arsenic** and most inorganic pollutants, can also be taken up through the root system and accumulate in grain or fruit. In addition to **abiotic** contaminants, soils can contain pathogenic fungi that produce **mycotoxins**, contaminating plant products and causing acute and chronic diseases in animals and humans. Furthermore, soils are also the source of **parasitic** worms (**helminthiasis**) that can live for years in the human **gastrointestinal** tract, cause malnutrition and result in stunted development.

⓭ The soil-health concept fills an important stakeholder need in sustainable development by elevating the recognition of the role of soil in modern society and is developing into an attractive and actionable platform for farmers, land managers, **municipalities**, and policymakers. The **versatility** of the concept allows many stakeholders to adopt soil health and to make it work for their context. By providing an illustrative link to broader sustainability goals that can motivate innovative soil management, soil health meets universal agreement in the eye of the public as a goal to work towards.

Unit 4 Soil Science

Note

(1) **US Department of Agriculture:** (informally the Agriculture Department or USDA) is the United States federal executive department responsible for developing and executing U.S. federal government policy on farming, agriculture, and food. It aims to meet the needs of farmers and ranchers, promote agricultural trade and production, work to assure food safety, protect natural resources, foster rural communities and end hunger in the United States and abroad.

Reading Comprehension

 Keys

Directions Match the paragraphs with their appropriate subtitles.

1. Paragraph 1 _____   A. Effects of soil on human health

2. Paragraphs 2–6 _____   B. Significance of the concept of soil health

3. Paragraph 7 _____   C. The definition of soil health

4. Paragraph 8 _____   D. Soil health management

5. Paragraphs 9 and 10 _____   E. Effects of soil health on plants

6. Paragraphs 11 and 12 _____   F. The challenges of defining "soil health"

7. Paragraph 13 _____   G. The origin and development of the concept of soil health

Exercises

  Keys

Vocabulary

I. Technical words

Directions Write out the English expressions according to the Chinese.

风化层　　　_____

腐殖土　　　_____

植被　　　　_____

苔原　　　　_____

真菌　　　　_____

原生动物　　_____

地形,地势　　_____

硝酸盐　　　_____

叶绿素　　　　　　_____

病原体　　　　　　_____

Ⅱ. Academic vocabulary

Directions　Fill in the blanks with appropriate forms of the words given below.

> designation　　decompose　　deposit　　compound　　sustainable
> evolve　　organic　　weather　　accumulate　　disintegrate

1. Bones _____, the land erodes, the climate changes, forests come and go, rivers change their course—and history, if not destroyed, is steadily concealed.
2. When animals or plants _____, they gradually change and develop into different forms.
3. When things such as dead plants or animals _____, or when something decomposes them, they change chemically and begin to decay.
4. Hurricane Katrina is widely considered the measure for a destructive storm, holding the maximum Category 5 _____ for a full 24 hours in late August 2005.
5. Leaves and tissue of soft-bodied organisms such as jellyfish or worms may _____, become buried and compressed, and lose their volatile constituents.
6. _____ of rocks is essentially a static process.
7. The versatile _____ is endostatin（内皮抑制素）, a human protein that inhibits angiogenesis, the growth of new blood vessels in the body.
8. The _____ belongs to a poly genetic compound gold deposit resulting from multi-mineralization and its formation is associated with multi-stage evolution of the crust in the area.
9. The bright plumage of many male birds was thought to have _____ to attract females.
10. Organized by the New York-based nonprofit Earth Pledge, the show inspired many top designers to work with _____ fabrics for the first time.

Sentence Structure

Understand object clauses in English sentences.

An object clause is a clause which serves as an object in a compound sentence. An object clause can function as a direct object, an indirect object, or the object of a preposition. Both statements and questions can function as object clauses. However, the words in an object clause should come in the statement order.

> For example,
> The goal of our company is *what is explained in the profile*. (direct object)
> Don't forget to send *whoever attended the conference* a follow-up email. (indirect object)
> You should speak about *how your parents immigrated to this country*. (object of a preposition)

Directions　Analyze the following sentences by: 1) underlining the object clauses, and 2) figuring out the type of each object clause.

1. Note that transported soils form in place on parent materials that have been carried from elsewhere and deposited by gravity, water, wind, or ice. (Paragraph 5, Passage A)
2. However, soil scientists came to understand that other factors, especially climate, are moreimportant. (Paragraph 7, Passage A)
3. Variations in temperature and precipitation determine whether chemical or mechanical weathering predominates. (Paragraph 9, Passage A)
4. Although we have dealt separately with each of the soil-forming factors, remember that all of them work together to form soil. (Paragraph 15, Passage A)

Translation

Ⅰ. **Translate the following English sentences into Chinese. Pay attention to the object clause in each sentence.**

1. Note that transported soils form in place on parent materials that have been carried from elsewhere and deposited by gravity, water, wind, or ice.

2. However, soil scientists came to understand that other factors, especially climate, are more important.

3. Variations in temperature and precipitation determine whether chemical or mechanical weathering predominates.

4. Although we have dealt separately with each of the soil-forming factors, remember that all of them work together to form soil.

Ⅱ. **Translate the following Chinese paragraph into English.**

虽然土壤是病原体的宿主,但历史上它也一直是生产链霉素等医疗行业使用的抗生素的微生物来源。大部分土壤微生物仍有待鉴定,人类医学应用的仍会有重要发现。土壤生物多样性的量化和管理是土壤健康管理目标的一部分,需要阻止微生物物种的灭绝,并为未来的生物勘探保存机会。

Unit 5 Modern Hydrology and Marine Sciences

Water is present in all parts of the Earth system, even in the rock of the Earth's interior. In other words, the hydrosphere overlaps with the geosphere, the atmosphere, the biosphere, and even the anthroposphere. It is mainly for convenience of discussion that we distinguish the hydrosphere as the "water sphere", for in it resides the bulk of the Earth's water in the surface and near-surface environment. In this unit, you will read about: 1) the hydrologic cycle, 2) the glacier formation, and 3) the ocean water respectively.

Passage A

flux *n.* 流量,流动
reservoir *n.* 水库,水箱,储液器
evaporation *n.* 蒸发
hydrologic cycle 水文循环,水循环
atmosphere *n.* 大气,大气层

The Hydrologic Cycle*

❶ The most familiar cycle of the Earth system is surely the hydrologic cycle[(1)], which describes the **fluxes** of water between the various **reservoirs** of the hydrosphere, the totality of the Earth's water on and just below the surface. We are familiar with these fluxes because we experience them as rain and snow or as a wet pavement drying by **evaporation**, and we see water moving and stored on the surface in streams, lakes, and wetlands. The movements of water through the **hydrologic cycle** and the important roles of water in the Earth system are the focus of this passage.

❷ Water in the atmosphere and large bodies of surface water play a central role in moderating temperature and controlling climate. They are the source of much of the water vapor in the **atmosphere**, and they store heat energy, exchanging it with the

---

* This passage is adapted from SKINNER B J, MURCK B W, 2011. The blue planet: an introduction to Earth system science [M]. 3rd ed. Hoboken, NJ: Wiley.

atmosphere. Another important consequence of the hydrologic cycle is the great diversity of the Earth's landscapes. The **erosional** and **depositional** effects of streams, waves, and glaciers, coupled with tectonic movements, have produced landscapes that make the Earth's surface unlike that of any other planet in the solar system. Through its effects on erosion and **sedimentation**, the hydrologic cycle is intimately related to the rock cycle. Finally, water is a key component of an array of biogeochemical cycles that control the composition of the atmosphere and influence all living creatures on the Earth.

❸ The unique properties of water as a chemical compound make life possible on this planet. Like all of the cycles in the Earth system, the hydrologic cycle is composed of pathways and reservoirs[2]. The pathways are the means by which water cycles between reservoirs. The reservoirs are the "storage tanks", where water may be held for varying lengths of time. The total amount of water in the hydrologic system is fixed, but there can be quite short-term large **fluctuations** in local reservoirs. For example, a river may flood in one area, while a drought occurs in an adjacent area. On a global scale, however, these local fluctuations do not change the total volume of water in the Earth system; the hydrologic cycle therefore can be said to maintain a mass balance on a global scale. Overall, it is a closed cycle, but within this closed cycle water is constantly shifting from one reservoir to another through a network of open subsystems.

❹ Although water is continuously moving from one reservoir to another, the volume of water in each reservoir is approximately constant over short time intervals. Over lengthy intervals, however, the volume of water in the different reservoirs can change dramatically. During glacial ages, for example, vast quantities of water were evaporated from the ocean and **precipitated** on land as snow. The snow slowly accumulated to build **ice sheets** that were thousands of meters thick and covered vast areas where none exists today. At the **culmination** of the most recent glacial age, the amount of water removed from the ocean was so large that sea level was about 120 m lower, and the expanded glaciers increased the ice-covered area of the Earth by more than 300 percent.

erosional *adj.* 侵蚀的,冲蚀的
depositional *adj.* 沉积作用的
sedimentation *n.* 沉积(作用)
fluctuation *n.* 波动,变动,起伏现象
precipitate *v.* 降水(如雨、雪、冰雹)
ice sheet 冰原,冰层,冰盖
culmination *n.* 终点,高潮

polar *adj.* 极地的,来自极地的
groundwater *n.* 地下水
pore space 孔隙
residence time 停留时间,滞留时间
ice cap 冰盖,冰原,冰冠
intercept *v.* 拦截,阻截
transpiration *n.* 蒸发,散发,蒸腾作用

❺ The largest reservoir for water in the hydrologic cycle is the ocean, which contains more than 97.5 percent of all the water in the Earth system. This means that most of the water in the hydrologic cycle is saline, not fresh. This has important implications for humans because we are dependent on fresh water as a resource for drinking, agriculture, and industrial uses. Surprisingly, the largest reservoir of fresh water is the great **polar** ice sheets, which contain almost 74 percent of all fresh water. The ice sheets are a long-term reservoir; water may be stored there for hundreds of thousands of years before it is recycled. Of the remaining unfrozen fresh water, almost 98.5 percent resides underground in the next largest reservoir, **groundwater**. Only a very small fraction of the water passing through the hydrologic cycle resides in surface freshwater bodies, such as streams and lakes. A smaller amount resides in **pore spaces** in soils, and an even smaller amount resides for short periods in the atmosphere and the bodies of living organisms.

❻ In general, a correlation exists between the size of a reservoir and the average time that water stays in that reservoir, its **residence time**. Residence time in the large-volume reservoirs, such as the ocean and **ice caps**, is many thousands of years, whereas in the small volume reservoirs it is short—a few days in the atmosphere, a few weeks in streams and rivers, a few days or hours in the bodies of living organisms.

❼ The movement of water among the Earth's reservoirs in the hydrologic cycle is powered by the Sun. Heat from the Sun causes evaporation of water from the ocean and land surfaces, in which water is converted from its liquid form to its vapor (gaseous) form. The water vapor thus produced enters the atmosphere and moves with the flowing air. With changing atmospheric conditions, some of the water vapor in the atmosphere undergoes condensation, changing from a vapor back into a liquid or solid state. The condensed water, gathering into droplets or particles, falls under the influence of gravity as precipitation (such as rain or snow) on the land or ocean. Rain falling on land may evaporate, or it may be **intercepted** by vegetation, subsequently returning to the atmosphere by evaporation from leaf surfaces. Plants also can return water to the atmosphere by **transpiration**.

❽ A reliable water supply is critical—not only for human survival and health, but also for the role it plays in industry, agriculture, and other economic activities, and for the environmental services it performs, such as supporting ecosystems, carrying away **contaminants**, and moderating climate. Water is under threat almost everywhere in the world, in terms of both quantity and quality.

❾ Today lots of countries worldwide suffer from significant water shortages; they are said to be experiencing water stress. The lack of water in these countries places constraints on agricultural production, economic development, health, and environmental protection. Sometimes regions with the greatest demand for water do not have an abundant and readily available supply of surface water.

❿ The accessibility of surface water bodies makes them useful as resources, but renders them highly susceptible to contamination. Contaminants in surface water come mainly from urban, suburban, and agricultural runoff. Industrial effluent (contaminated runoff) is a significant contributor to surface water pollution. So are discharges related to resource extraction, among which mining, logging, and the petroleum industry are important. Effluents from poorly engineered landfills are another source of surface water contamination.

⓫ The water that comes to our homes and offices typically is drawn from relatively clean surface water bodies or groundwater **aquifers**, but may still require treatment to meet drinking-water standards. Water quality improvements can include treatments for hardness, the addition of **fluoride** (fluoridation), and, where necessary, **desalination** to remove salts from the water. Desalination is energy-intensive and very expensive, but it is worth the cost in some countries where water is particularly scarce. Most importantly, water for domestic use must be filtered to remove sediment and other solids, and treated by **chlorination** to kill harmful microorganisms.

⓬ Ensuring the quality of water involves determining the health effects of toxins, and then setting and enforcing limits. An important aspect of this is to understand the pathways and processes by which toxic substances move through the various reservoirs of the Earth system and into organisms, including people.

contaminant n. 污染物,致污物
aquifer n. 地下含水层
fluoride n. 氟化物
desalination n. (海水的)脱盐,淡化
chlorination n. 氯化作用,加氯消毒

Notes

(1) **The hydrologic cycle or the hydrological cycle**, also known as the water cycle, is a biogeochemical cycle that describes the continuous movement of water on, above, and below the surface of the Earth. The hydrologic cycle interacts with the rock cycle, tectonic cycle, and biogeochemical cycles, and is interconnected with virtually all aspects of the Earth system.

(2) **Components of hydrologic cycle**: Evaporation, condensation, and precipitation are three major components of the water cycle.

Reading Comprehension

 Keys

Directions Match the paragraphs with their appropriate subtitles.

1. Paragraph 1 _____   A. Pathways in the hydrologic cycle
2. Paragraphs 2 – 4 _____   B. Water and the hydrologic cycle
3. Paragraphs 5 and 6 _____   C. Water and society
4. Paragraph 7 _____   D. Water shortages
5. Paragraph 8 _____   E. Reservoirs in the hydrologic cycle
6. Paragraph 9 _____   F. Water quality
7. Paragraph 10 _____   G. Definition of hydrologic cycle
8. Paragraphs 11 and 12 _____ H. Surface water contamination

Passage B

 Scan and read along the passage

Glacier Formation*

 Scan and read along the vocabulary

hexagonal *adj.* 六边的，六角形的
compacted *adj.* 压实的，压紧的

❶ Glaciers are slowly moving masses of ice that have accumulated on land in areas where more snow falls during a year than melts. Snow falls as **hexagonal** crystals, but once on the ground, snow is soon transformed into a **compacted** mass of

---

* This passage is adapted from KMF 考满分, 2015. Glacier formation[EB/OL]. [2022-05-10]. https://toefl.kmf.com/detail/read/01fdvj.html/1.

grains *n.* 谷粒
firn *n.* 积雪
interlocked *adj.* 连锁的
surplus *adj.* 过剩的,剩余的,多余的
    n. 过剩,剩余,过剩量
Antarctica *n.* 南极洲

smaller, rounded **grains**. As the air space around them is lessened by compaction and melting, the grains become denser. With further melting, refreezing, and increased weight from newer snowfall above, the snow reaches a granular recrystallized stage intermediate between flakes and ice known as **firn**. With additional time, pressure, and refrozen meltwater from above, the small firn granules become larger, **interlocked** crystals of blue glacial ice. When the ice is thick enough, usually over 30 meters, the weight of the snow and firn will cause the ice crystals toward the bottom to become plastic and to flow outward or downward from the area of snow accumulation.

❷ Glaciers are open systems, with snow as the system's input and meltwater as the system's main output. The glacial system[1] is governed by two basic climatic variables: precipitation and temperature. For a glacier to grow or maintain its mass, there must be sufficient snowfall to match or exceed the annual loss through melting, evaporation, and calving, which occurs when the glacier loses solid chunks as icebergs to the sea or to large lakes. If summer temperatures are high for too long, then all the snowfall from the previous winter will melt. **Surplus** snowfall is essential for a glacier to develop. A surplus allows snow to accumulate and for the pressure of snow accumulated over the years to transform buried snow into glacial ice with a depth great enough for the ice to flow. Glaciers are sometimes classified by temperature as faster-flowing temperate glaciers or as slower-flowing polar glaciers[2].

❸ Glaciers are part of the Earth's hydrologic cycle and are second only to the oceans in the total amount of water contained. About two percent of the Earth's water is currently frozen as ice. Two percent may be a deceiving figure, however, since over 80 percent of the world's fresh water is locked up as ice in glaciers, with the majority of it in **Antarctica**. The total amount of ice is even more awesome if we estimate the water released upon the hypothetical melting of the world's glaciers. Sea level would rise about 60 meters. This would change the geography of the planet considerably. In contrast, should another ice age occur, sea level would drop drastically. During the last ice age, sea level dropped about 120 meters.

❹ When snow falls on high mountains or in polar regions, it may become part of the glacial system. Unlike rain, which returns rapidly to the sea or atmosphere, the snow that becomes part of a glacier is involved in a much more slowly cycling system. Here water may be stored in ice form for hundreds or even hundreds of thousands of years before being released again into the liquid water system as meltwater. In the meantime, however, this ice is not static. Glaciers move slowly across the land with tremendous energy, carving into even the hardest rock formations and thereby reshaping the landscape as they **engulf**, push, drag, and finally deposit rock debris in places far from its original location. As a result, glaciers create a great variety of landforms that remain long after the surface is released from its icy covering.

engulf *v.* 吞没,淹没
elevations *n.* 高度,海拔
periodically *adv.* 定期地,周期性地

❺ Throughout most of the Earth's history, glaciers did not exist, but at the present time about 10 percent of the Earth's land surface is covered by glaciers. Present-day glaciers are found in Antarctica, in Greenland, and at high **elevations** on all the continents except Australia. In the recent past, from about 2.4 million to about 10,000 years ago, nearly a third of the Earth's land area was **periodically** covered by ice thousands of meters thick. In the much more distant past, other ice ages have occurred.

Notes

(1) **The glacial system** consists of inputs, stores, transfers, and outputs. Understanding the connections and relationships between the different components of the system helps indicate how a change in one component results in consequential changes throughout the system.

(2) **A polar glacier** is defined as one that is below the freezing temperature throughout its mass for the entire year.

Reading Comprehension

Keys

Directions Answer the following questions according to Passage B.

1. The word "interlocked" in Paragraph 1 is closest in meaning to ( )
   A. intermediate.
   B. linked.
   C. frozen.
   D. fully developed.
2. According to Paragraph 1, which of the following does NOT describe a stage in the development of firn? ( )

A. Hexagonal crystals become larger and interlock to form a thick layer.

B. Snow crystals become compacted into grains.

C. Granules recrystallize after melting, refreezing, and further compaction.

D. Grains become denser owing to reduced air space around them.

3. According to Paragraph 2, surplus snow affects a glacier in all the following ways EXCEPT ( )

    A. it provides the pressure needed to cause glacial ice to flow.

    B. it offsets losses of ice due to melting, evaporation, and calving.

    C. it brings about the formation of firn in the snow it buries.

    D. it results in temperate glaciers that are thicker than polar glaciers.

4. Paragraph 2 implies that which of the following conditions produces the fastest moving glaciers? ( )

    A. A climate characteristic of the polar regions.

    B. A thick layer of ice in a temperate climate.

    C. Long, warm summers.

    D. Snow, firn, and ice that have been buried for several years.

5. Why does the author consider "the hypothetical melting of the world's glaciers"? ( )

    A. To contrast the effects of this event with the opposite effects of a new ice age.

    B. To emphasize how much water is frozen in glaciers.

    C. To illustrate the disastrous effects of a warming trend.

    D. To support the claim that glaciers are part of the Earth's hydrologic cycle.

6. The discussion in Paragraph 3 answers all the following questions EXCEPT ( )

    A. Where is most of the Earth's freshwater?

    B. What effect would a new ice age have on sea levels?

    C. What is the total amount of water in the Earth's oceans?

    D. How much of the Earth's water is in ice?

7. Which of the sentences below best expresses the essential information in the highlighted sentence in the passage? ( )

    A. As a glacier moves, it leaves behind rock formations that have been engulfed, pushed, and dragged by the glacier.

    B. Glaciers reshape the landscape by carving into rock and transporting the resulting debris to distant locations.

    C. Glaciers carve the hardest rock formations with great energy and slowly reshape them into debris.

    D. The tremendous energy of slowly moving glaciers transports and finally deposits rock debris into large rock formations.

8. According to Paragraph 5, in what way is the present time unusual in the history of the Earth? ( )

    A. There are glaciers.

B. More land is covered by glaciers than at any time in the past.

C. There is no ice age.

D. No glaciers are found in Australia.

9. The letters [A], [B], [C], and [D] indicate where the following sentence (in bold) could be added to the following part of the passage. Where would the sentence best fit?　　(　)

Firn has the appearance of wet sugar, but it is almost as hard as ice.

[A] As the air space around them is lessened by compaction and melting, the grains become denser. [B] With further melting, refreezing, and increased weight from newer snowfall above, the snow reaches a granular recrystallized stage intermediate between flakes and ice known as firn. [C] With additional time, pressure, and refrozen meltwater from above, the small firn granules become larger, interlocked crystals of blue glacial ice. [D] When the ice is thick enough, usually over 30 meters, the weight of the snow and firn will cause the ice crystals toward the bottom to become plastic and to flow outward or downward from the area of snow accumulation.

10. An introductory sentence (in bold) for a brief summary of the passage is provided below. Complete the summary by selecting the THREE answer choices that express the most important ideas in the passage.　　(　)

Glaciers are part of the Earth's hydrologic cycle.

A. Glaciers, which at present contain 80 percent of the Earth's freshwater, form when accumulated snow is compressed and recrystallized into ice over a period of years.

B. When glacial ice reaches a depth of 30 meters, the weight of the ice causes ice crystals at the bottom to flow, and the resulting movement of the glacier carves the landscape.

C. When there are glaciers on the Earth, water is cycled through the glacier system, but the cycle period may be hundreds of thousands of years during periods of ice ages.

D. If global warming melted the world's glaciers, sea level would rise about 60 meters worldwide.

E. The glacial system is governed by precipitation and temperature in such a way that glaciers cannot form in temperate latitudes.

F. Glaciers have had little effect on the Earth's surface because only 2 percent of the Earth's water is currently contained in glaciers, and there are fewer glaciers now than at most times in the past.

Passage C *Scan and read along the passage*

Ocean Water*

 Scan and read along the vocabulary

❶ Imagine yourself lying on a beach and listening to the waves gently roll onto shore. A warm breeze blows off the water, making it seem as if you're in a tropical paradise. It's easy to appreciate the oceans under these circumstances, but the oceans affect your life in other ways, too.

❷ Oceans are important sources of food, energy, and minerals. Energy sources such as oil and natural gas are found beneath the ocean floor. Oil wells are often drilled in shallow water. Mineral resources including copper and gold are mined in shallow waters as well. Approximately one third of the world's table salt is extracted from seawater through the process of evaporation. Oceans also allow for the efficient transportation of goods. For example, millions of tons of oil, coal, and grains are shipped over the oceans each year.

❸ During the Earth's first billion years, its surface was much more volcanically active than it is today. When volcanoes erupt, they **spew** lava and ash, and they give off water vapor, **carbon dioxide**, and other gases. Scientists hypothesize that about 4 billion years ago, this water vapor began to be stored in the Earth's early atmosphere. Over millions of years, it cooled enough to condense into storm clouds. **Torrential** rains began to fall, and oceans were formed as this water filled low areas on the Earth called **basins**. Today, approximately 70 percent[1] of the Earth's surface is covered by ocean water.

❹ Ocean water contains **dissolved** gases such as oxygen, carbon dioxide, and nitrogen. Oxygen is the gas that almost all organisms need for **respiration**. It enters the oceans in two ways—directly from the atmosphere and from **organisms** that **photosynthesize**. Carbon dioxide enters the ocean from the atmosphere and from organisms when they respire. The atmosphere is the only

spew *v.* (使)喷出,呕吐
carbon dioxide *n.* 二氧化碳
torrential *adj.* (雨)倾泻的,如注的
basin *n.* 盆地,海盆
dissolved *adj.* 溶解的,溶化的
respiration *n.* 呼吸
organism *n.* 生物,有机体
photosynthesize *v.* 起光合作用,光能合成

---

* This passage is excerpted from FEATHER R M, SNYDER S L, ZICK D, 1993. Earth science[M]. Carson City, NV: Merrill Publishing Company.

bacteria *n.* 细菌（bacterium 的复数）
sulfate *n.* 硫酸盐
magnesium *n.* （化学元素）镁
atom *n.* 原子
bromide *n.* 溴化物
hydrogen *n.* 氢，氢气
halite *n.* 岩盐

important source of nitrogen gas. **Bacteria** combine nitrogen and oxygen to create nitrates, which are important nutrients for plants.

❺ If you've ever tasted ocean water, you know that it is salty. Ocean water contains many dissolved salts. Chloride, sodium, **sulfate, magnesium**, calcium, and potassium are some of the ions in seawater. An ion is a charged **atom** or group of atoms. Some of these ions come from rocks that are dissolved slowly by rivers and groundwater. These include calcium, magnesium, and sodium. Rivers carry these chemicals to the oceans. Erupting volcanoes add other ions, such as **bromide** and chloride.

❻ The most abundant elements in sea water are the **hydrogen** and oxygen that make up water. Many other ions are found dissolved in seawater. When seawater is evaporated, these ions combine to form materials called salts. Sodium and chloride make up most of the ions in seawater. If seawater evaporates, the sodium and chloride ions combine to form a salt called halite. **Halite** is the common table salt you use to season food. It is this dissolved salt and similar ones that give ocean water its salty taste.

❼ It is a measure of the amount of salts dissolved in seawater. It is usually measured in grams (g) of dissolved salt per kilogram (kg) of seawater. One kilogram of ocean water contains about 35 g of dissolved salts, or 3.5 percent. Ocean water contains about 96.5 percent pure water and 3.5 percent dissolved salts. There are varied icons contained in the dissolved salts: 55 percent Chloride, 30.6 percent sodium, 7.7 percent sulfate, 3.7 percent magnesium, 1.2 percent calcium, 1.1 percent potassium, and 0.7 other. The proportion and amount of dissolved salts in seawater remain nearly constant and have stayed about the same for hundreds of millions of years. This tells you that the composition of the oceans is in balance. Evidence that scientists have gathered indicates that the Earth's oceans are not growing saltier.

❽ Although rivers, volcanoes, and the atmosphere constantly add material to the oceans, the oceans are considered to be in a steady state. This means that elements are added to the oceans at about the same rate that they are removed. Dissolved salts are removed when they precipitate out of ocean water and become part of the sediment. Some marine organisms use dissolved salts to make body parts. Some remove calcium ions from the water to form bones. Other animals, such as **oysters**, use the dissolved calcium to form shells. Some **algae**, called **diatoms**, have **silica** shells. Because many organisms use calcium and silicon, these elements are removed more quickly from seawater than elements such as chlorine or sodium.

❾ Salt can be removed from ocean water by a process called desalination. If you have ever swum in the ocean, you know what happens when your skin dries. The white, flaky substance on your skin is salt. As seawater evaporates, salt is left behind. As demand for freshwater increases throughout the world, scientists are working on technology to remove salt to make seawater drinkable.

❿ Many varieties of plants and animals live in the salty ocean. Although some organisms live in the open ocean or on the deep ocean floor, most marine organisms live in the waters above or on the floor of the continental shelf. In this relatively shallow water, the Sun penetrates to the bottom, allowing for photosynthesis. Because light is available for photosynthesis, large numbers of producers live in the waters above the **continental shelf**[2]. These waters also contain many nutrients that producers use to carry out life processes. As a result, the greatest source of food is located in the waters of the continental shelf.

⓫ The plants and animals living on or in the seafloor are the **benthos**. Benthic animals include crabs, snails, **sea urchins**, and **bottom-dwelling** fish such as **flounder**. These organisms move or swim across the bottom to find food. Other benthic animals that live permanently attached to the bottom, such as sea **anemones** and sponges, filter out food particles from seawater. Certain types of worms live burrowed in the sediment of the ocean floor. Bottom-dwelling animals can be found living from the shallow water of the continental shelf to the deepest areas of the ocean. Benthic plants and algae, however, are limited to the shallow areas of the ocean where enough sunlight penetrates the water to allow for photosynthesis. One example of a benthic algae is kelp, which is anchored to the bottom and grows toward the surface from depths of up to 30 m.

oyster *n.* 牡蛎,蚝
algae *n.* 水藻,海藻
diatom *n.* 硅藻
silica *n.* 二氧化硅,硅土
continental shelf *n.* 大陆架
benthos *n.* 海底生物,海底的动植物群
sea urchin 海胆
bottom-dwelling *adj.* 底栖的
flounder *n.* 比目鱼(同 flatfish)
anemone *n.* 银莲花,海葵(sea anemone)

Notes

(1) **More than 97 percent** of the Earth's water resides in the ocean. The ocean dominates the surface of our planet and plays many crucial roles in influencing climate and supporting life.

(2) **A continental shelf** is a portion of a continent that is submerged under an area of relatively shallow water known as a shelf sea. Much of these shelves were exposed by drops in sea level during glacial periods. The shelf surrounding an island is known as an insular shelf.

Reading Comprehension

Directions   Match the paragraphs with their appropriate subtitles.

1. Paragraphs 1 and 2 ____
2. Paragraph 3 ____
3. Paragraphs 4 and 5 ____
4. Paragraphs 6 and 7 ____
5. Paragraph 8 ____
6. Paragraph 9 ____
7. Paragraph 10 ____
8. Paragraph 11 ____

A. Origin of oceans
B. Importance and varied resources of oceans
C. Ocean life
D. Desalination
E. Removal of elements
F. Composition of oceans
G. Bottom dwellers
H. Salts and salinity

Exercises

Vocabulary

Ⅰ. Technical words

Directions   Write out the English expressions according to the Chinese.

水圈　　　　　　　　　　_____

水循环　　　　　　　　　_____

地下水　　　　　　　　　_____

冰冻圈　　　　　　　　　_____

冰盖,冰原　　　　　　　_____

降水　　　　　　　　　　_____

(海水的)脱盐　　　　　 _____

海底生物　　　　　　　　_____

Unit 5 Modern Hydrology and Marine Sciences

I. Academic vocabulary

Directions Fill in the blanks with appropriate forms of the words given below. Each word can be used only once.

> describe   consequence   effect   implication   reside
> govern   periodically   imagine   constantly   hypothesize

1. One of the more interesting and perhaps dramatic _____ of the Ice Age was the fall and rise of sea level that accompanied the advance and retreat of the glaciers.
2. Scientists _____ that organic molecules (or possibly even life, itself) may have arrived, ready made, from some other part of the solar system or even from the galaxy beyond the solar system.
3. All of the impacts _____ above can have negative impacts on plants and animals, including humans.
4. Comets are small, loosely packed, icy bodies—like dirty snowballs. They travel _____ to the inner part of the solar system, following highly elongate, elliptical orbits.
5. There is evidence that dramatic, rapid changes of this type have occurred in the geologic past; this evidence, the causes, and _____ of such changes will be discussed in this Chapter.
6. Water is _____ moving among the Earth's different spheres—the hydrosphere, the atmosphere, the geosphere, and the biosphere.
7. The laws of thermodynamics _____ the behavior of energy as it moves through the Earth system, into and out of the Earth system, and among its reservoirs.
8. A very small fraction of the water in the hydrosphere _____ in surface freshwater bodies, soil moisture, and living organisms.
9. Convective heat transfer has significant _____ for life on the Earth's surface.
10. Then try to _____ an earth with many smaller ocean basins surrounding smaller, island-like continents. What would be the impacts on climate and on ecosystems? What would it be like to live on such an earth?

Sentence Structure

Understand noun clauses in English sentences.

A noun clause is a dependent (or subordinate clause) that works as a noun. It can be the subject, an object, or a complement of a sentence. Like all nouns, the purpose of a noun clause is to name a person, place, thing, or idea. Sometimes a single word is not enough to name something. Instead, a group of words are needed to name something. That is why noun clauses are used.

Types of Noun Clauses:

Noun clauses as subjects.

For example,
  Whoever leaves last should turn off the lights.
  In music, *which note is played and how long it is played* are both essential.

Noun clauses as objects. There are three types of objects, as shown below.

(1) Direct objects—receive the action of the verb.

For example,
  My dog will eat *whatever food I give him.*

(2) Indirect objects—receive direct objects.

For example,
  The judges will award *whichever painting they like the most* the blue ribbon.

(3) Objects of prepositions—receive prepositions.

For example,
  I want to play with *whoever is a good sport.*

Noun clauses as compliments. A compliment re-states or gives more information about a noun. It always follows a state-of-being verb (is, are, am, will be, was, were).

For example,
  The winner will be *whoever gets the most votes.*
  My hope is *that everyone here becomes friends.*

Directions Analyze the following sentences by: 1) underlining the noun clause(s), and 2) identifying the type of each noun clause.

1. This means that most of the water in the hydrologic cycle is saline, not fresh. (Paragraph 5, Passage A)

2. Of particular concern is that some contaminants have long residence times in natural reservoirs; they are said to be persistent. (Paragraph 12, Passage A)

3. Scientists hypothesize that about 4 billion years ago, this water vapor began to be stored in the Earth's early atmosphere. (Paragraph 3, Passage C)

4. Evidence that scientists have gathered indicates that the Earth's oceans are not growing saltier. (Paragraph 7, Passage C)

Translation

Ⅰ. Translate the following English sentences into Chinese. Pay attention to the noun clause in each sentence.

1. This means that most of the water in the hydrologic cycle is saline, not fresh.

2. Of particular concern is that some contaminants have long residence times in natural reservoirs; they are said to be persistent.

3. Scientists hypothesize that about 4 billion years ago, this water vapor began to be stored in the Earth's early atmosphere.

4. Evidence that scientists have gathered indicates that the Earth's oceans are not growing saltier.

Ⅱ. Translate the following Chinese paragraph into English.

地下水是指渗入地下并填满岩石空隙的水。迄今为止最丰富的地下水来源便是大气水,它是自然界水循环的一部分。普通大气水会从地表、雨雪湖泊和河川渗入地下。有时它们会长时间滞留于此,之后再次涌出地表。或许在人们看来,坚硬的地壳居然有充足空间来储存如此多的水是难以置信的。然而,地下水的储存形式是多种多样的。

Unit 6 Resources and Energy on the Earth

Natural resources are essential to the survival of humans and all other living organisms. All the products in the world use natural resources as their basic component, which may be water, air, natural chemicals or energy. In this unit, you will get to know: 1) natural resources, including its classification, protection, and sustainable development; 2) the energy cycle of the Earth system; and 3) one of the new energy resources—nuclear energy.

Passage A

Scan and read along the passage

Scan and read along the vocabulary

substance *n.* 物质，材料

Natural Resources: Classification and Protection*

❶ Natural resources are resources that are drawn from nature and used to support life and meet people's needs. Any natural **substance** that humans use can be considered a natural resource. Oil, coal, natural gas, metals, stone, and sand are natural resources. Other natural resources are air, sunlight, soil, and water. Animals, birds, fish, and plants are natural resources as well. Natural resources are used to make food, fuel, and raw materials for the production of goods. All of the food that people eat comes from plants or animals. Natural resources such as coal, natural gas, and oil provide heat, light, and power. Natural resources are also the raw materials for making products that we use every day from our toothbrush and lunch box to our clothes, cars, televisions, computers, and refrigerators.

---

* The passage is adapted from: 1) Natural resource[EB/OL]. (2022-06-13)[2022-06-27]. https://en.wikipedia.org/wiki/Natural_resource; 2) SAWE B E. What are natural resources?[EB/OL]. (2018-08-27)[2022-06-27]. https://www.worldatlas.com/articles/what-are-natural-resources.html.

❷ There are different criteria of classifying natural resources. These include the source of origin, level of development, **renewability**, etc. On the basis of origin, we could divide the various resources into biotic and abiotic ones. The word "bio" means life. Biotic resources are resources that originate from the biosphere and have life such as **flora** and **fauna, fisheries, livestock**, etc. Fossil fuels such as coal and petroleum are also included in this category because they are formed from decayed organic matter. Abiotic resources are those resources that originate from non-living or **inorganic** materials and have no life in them. Examples include land, fresh water, air, rare-earth elements, and heavy metals including **ores**, such as gold, iron, copper, silver, etc.

❸ Considering their stage of development, natural resources may also be divided into potential resources, actual resources, reserves, and stocks. Potential resources are those that are known to exist, but have not been **utilized** yet. For instance, wind energy exists in certain areas but has not been used to generate energy. Nevertheless, it could be used in the future. In this sense, petroleum in sedimentary rocks remains a potential resource unless it has been actually **drilled** out and put into use. Actual resources are resources that have been **surveyed**, quantified and qualified, and are currently used in development. The development of an actual resource, such as wood processing depends upon the technology available and the cost involved. Reserve resources are resources that have been identified and quantified but have not been **harnessed** because they are being reserved for future use. Stock resources are also resources that have been discovered, quantified and have not yet been harnessed, but due to insufficient technologies instead of reserve purposes. For example, we know that water consists of hydrogen and oxygen. Hydrogen is considered as one of the sources of energy but we do not know the technology to **extract** energy from it. We also lack the skills and technology to extract and use some of the naturally occurring resources like rare gases and some **radioactive** materials. These resources are, thus, classified as stock resources to be utilized in the future.

❹ Renewability is another very popular topic when talked of natural resources. Based on renewability or **exhaustibility**, natural

---

renewability n. 可再生性
flora n. (某地区、环境或时期的)植物群,植物界
fauna n. (某地区、环境或时期的)动物群,动物界
fishery n. 渔业,渔场,水产业
livestock n. 牲畜,家畜
inorganic adj. 无机的,无生物的
ore n. 矿石,矿砂
utilize v. 利用,使用
drill v. 钻(孔),打(眼)
survey v. 测量,勘测
harness v. 控制,利用(以产生能量等)
extract v. 提取,提炼
radioactive adj. 放射性的,有辐射的
exhaustibility n. 可用尽,耗竭性

non-renewable *adj.* 不可再生的，不可更新的
replenish *v.* 补充，重新装满，补足（原有的量）
uranium *n.* （放射性化学元素）铀
metallic *adj.* 含金属的，金属制的
sustainable *adj.* （自然资源）可持续的，不破坏环境的
sustainable development 可持续发展
inequity *n.* 不公平，不公正
ecosystem *n.* 生态系统

resources can be categorized as either renewable or **non-renewable**. Renewable resources can be **replenished** naturally. Some of these resources such as sunlight, air, and wind are continuously available, and their quantities are not noticeably affected by human consumption. Though many renewable resources do not have such a rapid recovery rate, and they are susceptible to depletion by over use, they are classified as renewable from a human use perspective only so long as the rate of replenishment or recovery exceeds that of consumption. Non-renewable resources formed over a long geological time period in the environment and cannot be renewed easily. Minerals are the most common resource included in this category. A good example of this is fossil fuels, which are in this category because their rate of formation is extremely slow (potentially millions of years), meaning they are considered non-renewable. Some resources naturally deplete in amount without human interference, the most notable of these being radio active elements such as **uranium**, which naturally decay into heavy metals. Of these, the **metallic** minerals can be re-used by recycling them, but coal and petroleum cannot be recycled. Once they are completely used they take millions of years to replenish.

❺ In recent years, the depletion of natural resources has become a major focus of governments and organizations such as the United Nations (UN). This is evident in the UN's Agenda 21[1] Section 2, which outlines the necessary steps for countries to take to sustain their natural resources. The depletion of natural resources is considered a **sustainable development** issue. The term "sustainable development" has many interpretations, most notably the Brundtland Commission's[2] "to ensure that it meets the needs of the present without compromising the ability of future generations to meet their own needs"; however, in broad terms it is balancing the needs of the planet's people and species now and in the future. In regards to natural resources, depletion is of concern for sustainable development as it has the ability to degrade current environments and the potential to impact the needs of future generations.

❻ Depletion of natural resources is associated with social **inequity**. Considering that most biodiversity are located in developing countries, depletion of this resource could result in losses of **ecosystem** services for these countries. Some view this depletion as a major source of social **unrest** and conflicts in developing nations.

❼ At present, there is a particular concern for rainforest regions that hold most of the Earth's biodiversity. According to Nelson, **deforestation** and **degradation** affect 8.5% of the world's forests with 30% of the Earth's surface already **cropped**. If we consider that 80% of people rely on medicines obtained from plants and 75% of the world's **prescription** medicines have ingredients taken from plants, loss of the world's rainforests could result in a loss of finding more potential life-saving medicines.

❽ The depletion of natural resources is caused by "direct **drivers** of change" such as mining, petroleum extraction, fishing, and forestry as well as "indirect drivers of change" such as **demography**, economy, society, politics, and technology. The current practice of agriculture is another factor causing depletion of natural resources. For example, the depletion of **nutrients** in the soil could be attributed to excessive use of nitrogen and **desertification**. The depletion of natural resources is a continuing concern for society. This is seen in the cited quote given by Theodore Roosevelt, a well-known **conservationist** and former United States president, who was opposed to **unregulated** natural resource extraction.

*The conservation of natural resources is the fundamental problem. Unless we solve that problem, it will **avail** us little to solve all others.*

Theodore Roosevelt

❾ In 1982, the United Nations developed the *World Charter for Nature*[3], which recognized the need to protect nature from further depletion due to human activity. It states that measures must be taken at all societal levels, from international to individual, to protect nature. It outlines the need for sustainable use of natural resources and suggests that the protection of resources should be incorporated into national and international systems of law. Other organizations like the International Union for Conservation of Nature (IUCN)[4] and the World Wide Fund for Nature (WWF)[5] have also led in the push for protection of natural resources. The organizations have funded scientific studies

---

unrest *n.* 不安,动荡的局面,不安的状态
deforestation *n.* 毁林,滥伐森林
degradation *n.* 退化,恶化,质量下降
crop *v.* 出露,露头
prescription *n.* 处方,药方
driver *n.* 驱动程序,驱动因素
demography *n.* 人口组成,人口统计
nutrient *n.* 养分,营养物
　　adj. 营养的,滋养的
desertification *n.* (土壤)荒漠化,沙漠化
conservationist *n.* (自然环境、野生动植物等)保护主义者
unregulated *adj.* 未受控制的,无管理的,未经调节的
avail *v.* 有帮助,有益,有用

conservation biology 保护生物学 exploitation n. 开发,开采,(出于私利、不公正的)利用 oversee v. 监管,监督	like **conservation biology** where scientists research on ways to conserve the natural resources found in the environment. At the local level, countries have established protected areas to conserve natural resources from **exploitation**. Conservationists also encourage the use of renewable natural resources such as wind and solar energy instead of non-renewable resources which are at risk of extinction. Additionally, most countries have government departments that **oversee** the extraction and use of natural resources. These departments create rules on management of natural resources like precious metals, rare metals, and energy sources. They also provide licenses to companies involved in the production and sale of such resources.

Notes

(1) **UN's Agenda 21** is a non-binding action plan of the United Nations with regard to sustainable development. It is a product of the Earth Summit (UN Conference on Environment and Development) held in Rio de Janeiro, Brazil, in 1992. It is an action agenda for the UN, other multilateral organizations, and individual governments around the world that can be executed at local, national, and global levels. One major objective of the Agenda 21 initiative is that every local government should draw its own local Agenda 21. Its aim initially was to achieve global sustainable development by 2000, with the "21" in "Agenda 21" referring to the original target of the 21st century.

(2) **The Brundtland Commission**, formerly the World Commission on Environment and Development, was a sub-organization of the United Nations that aimed to unite countries in pursuit of sustainable development. It was founded in 1983 when Gro Harlem Brundtland, the former Prime Minister of Norway, was appointed as chairperson of the commission. Brundtland was chosen due to her strong background in the sciences and public health.

(3) **The World Charter for Nature** was adopted by United Nations member nation-states on October 28, 1982. It proclaims five "principles of conservation by which all human conduct affecting nature is to be guided and judged": 1) nature shall be respected and its essential processes shall not be impaired; 2) the genetic viability on the Earth shall not be compromised; the population levels of all life forms, wild and domesticated, must be at least sufficient for their survival, and to this end necessary habitats shall be safeguarded; 3) all areas of the Earth, both land and sea, shall be subject to these principles of conservation; special protection shall be given to unique areas, to representative samples of all the different types of ecosystems and to the habitats of rare or endangered species; 4) ecosystems and organisms, as well as the land, marine, and atmospheric resources that are utilized by man, shall be managed to achieve and maintain optimum sustainable productivity, but not in such a

way as to endanger the integrity of those other ecosystems or species with which they coexist; and 5) nature shall be secured against degradation caused by warfare or other hostile activities.

(4) **International Union for Conservation of Nature (IUCN)** is an international organization working in the field of nature conservation and sustainable use of natural resources. It is involved in data gathering and analysis, research, field projects, advocacy, and education. IUCN's mission is to "influence, encourage and assist societies throughout the world to conserve nature and to ensure that any use of natural resources is equitable and ecologically sustainable".

(5) **World Wide Fund for Nature (WWF)** is an international non-governmental organization founded in 1961 that works in the field of wilderness preservation and the reduction of human impact on the environment. It was formerly named the World Wildlife Fund, which remains its official name in Canada and the United States.

Reading Comprehension

 Keys

Directions Fill in blanks according to Passage A.

Renewable Resource	Non-Renewable Resource
It can be renewed as it is available in 1._____ quantity.	Once completely consumed, it cannot be renewed due to 2._____ stock.
3._____ in nature.	4._____ in nature.
Low cost and environment-friendly.	5._____ cost and less environment-friendly.
Replenish 6._____.	Replenish slowly or do not replenish naturally at all.

Passage B

 Scan and read along the passage

The Earth's Energy Cycle*

❶ To understand most of the processes at work on the Earth, it is useful to **envisage** interactions within the Earth system as a series of **interrelated** cycles. One of these is the energy cycle[(1)], which encompasses the great "engines"—the external and internal energy sources—that drive the Earth system and all its cycles. We

 Scan and read along the vocabulary

envisage *v.* 想象,设想,展望
interrelated *adj.* 相关的,互相联系的

---

* This passage is adapted from 知乎专栏,2019. 备战托福:2019 年 11 月 16&17 日托福考试机经[EB/OL]. (2019-11-11) [2022-06-27]. https://zhuanlan.zhihu.com/p/91257082.

subtract v. 减去,删减,扣除
transfer v. 转移,搬迁,转移
terawatt n. (电功率单位)太(拉)瓦,万亿瓦
dwarf v. 使显得矮小,使相形见绌
overwhelmingly adv. 压倒性地,不可抵抗地
influx n. (人或物的)大量涌入,大量流入,注入
photosynthesis n. 光合作用
geothermal adj. 地热的,地温的
uplift v. 抬起,举起
n. (地壳的)隆起,举起,抬起
kinetic adj. 运动的,活跃的
bulge n. 鼓起,凸出,骤增,暴涨
axis n. 轴,轴线,对称中心线
rotate v. 旋转,转动
land mass 陆块,地块,大陆块

can think of the Earth's energy cycle as a "budget": energy may be added to or **subtracted** from the budget and may be **transferred** from one storage place to another, but overall the additions and subtractions and transfers must balance each other. If a balance did not exist, the Earth would either heat up or cool down until a balance was reached.

❷ The total amount of energy flowing into the Earth's energy budget is more than 174,000 **terawatts** (or $174,000 \times 10^{12}$ watts). This quantity completely **dwarfs** the 10 terawatts of energy that humans use per year. There are three main sources from which energy flows into the Earth system.

❸ Incoming short-wavelength solar radiation **overwhelmingly** dominates the flow of energy in the Earth's energy budget, accounting for about 99.986 percent of the total. An estimated 174,000 terawatts of solar radiation is intercepted by the Earth. Some of this vast **influx** powers the winds, rainfall, ocean currents, waves, and other processes in the hydrologic (or water) cycle. Some is used for **photosynthesis**[2] and is temporarily stored in the biosphere in the form of plant and animal life. When plants die and are buried, some of the solar energy is stored in rocks; when we burn coal, oil, or natural gas, we release stored solar energy.

❹ The second most powerful source of energy, at 23 terawatts or 0.013 percent of the total, is **geothermal** energy, the Earth's internal heat energy. Geothermal energy eventually finds its way to the Earth's surface, primarily via volcanic pathways. It drives the rock cycle and is therefore the source of the energy that **uplifts** mountains, causes earthquakes and volcanic eruptions, and generally shapes the face of the Earth.

❺ The smallest source of energy for the Earth is the **kinetic** (motion) energy of the Earth's rotation. The Moon's gravitational pull lifts a tidal **bulge** in the ocean; as the Earth spins on its **axis**, this bulge remains essentially stationary. As the Earth **rotates**, the tidal bulge runs into the coastlines of continents and islands, causing high tides. The force of the tidal bulge "piling up" against **land masses** acts as a very slow brake, actually causing the Earth's rate of rotation to decrease slightly. The transfer of tidal energy accounts for approximately 3 terawatts, or 0.002 percent of the tidal energy budget.

❻ The Earth loses energy from the cycle in two main ways: **reflection**, and degradation and **reradiation**. About 40 percent of incoming solar radiation is simply reflected, unchanged, back into space by the clouds, the sea, and other surfaces. For any planetary body, the percentage of incoming radiation that is reflected is called the "albedo". Each different material has a characteristic reflectivity. For example, ice is more reflectant than rocks or pavement; water is more highly reflectant than vegetation; and forested land reflects light differently than agricultural land. Thus, if large **expanses** of land are converted from forest to **plowed land**, or from forest to city, the actual reflectivity of the Earth's surface, and hence its albedo, may be altered. Any change in albedo will, of course, have an effect on the Earth's energy budget.

❼ The portion of incoming solar energy that is not reflected back into space, along with tidal and geothermal energy, is absorbed by materials at the Earth's surface, in particular the atmosphere and hydrosphere. This energy undergoes a series of **irreversible** degradations in which it is transferred from one reservoir to another and converted from one form to another. The energy that is absorbed, utilized, transferred, and degraded eventually ends up as heat, in which form it is reradiated back into space as long-wavelength (**infrared**) radiation. Weather patterns are a manifestation of energy transfer and degradation.

reflection n. （光、热或声音的）反射
reradiation n. 再辐射
expanse n. 宽阔，宽阔的区域，膨胀扩张
plowed land 耕地
irreversible adj. 不可逆的，不能取消的，不能翻转的
infrared adj. 红外线的

Notes

(1) **Energy Cycle:** Energy from the Sun is the driver of many the Earth system processes. This energy flows into the Atmosphere and heats this system up. It also heats up the hydrosphere and the land surface of the geosphere, and fuels many processes in the biosphere. Differences in the amount of energy absorbed in different places set the Atmosphere and oceans in motion and help determine their overall temperature and chemical structure. These motions, such as wind patterns and ocean currents redistribute energy throughout the environment. Eventually, the energy that began as sunshine (short-wave radiation) leaves the planet as earthshine (light reflected by the atmosphere and surface back into space) and infrared radiation (heat, also called long-wave radiation) emitted by all parts of the planet which reaches the top of the Atmosphere. This flow of energy from the Sun, through the environment, and back into space is a major connection in the Earth system; it defines the

Earth's climate.

(2) **Photosynthesis** is the process by which green plants and certain other organisms transform light energy into chemical energy. During photosynthesis in green plants, light energy is captured and used to convert water, carbon dioxide, and minerals into oxygen and energy-rich organic compounds.

Reading Comprehension

  Keys

Directions Answer the following questions according to Passage B.

1. In Paragraph 1, the author introduces the concept of a "budget" in order to (   )
   A. indicate how different cycles in the Earth system relate to each other.
   B. illustrate how the Earth's energy cycle must maintain an overall balance.
   C. show that the Earth gains energy from both external and internal sources.
   D. explain how energy is transferred from one storage place to another.

2. In Paragraph 2, why does the author include information about the "energy that humans use per year" in the discussion? (   )
   A. To call into question the idea that humans can use up all the energy available in the Earth's energy budget.
   B. To provide a comparison that established how huge the amount of energy flowing into the Earth's energy budget is.
   C. To explain why there must be more than one source of energy for the Earth system.
   D. To argue that the use of energy by humans amounts to such a small part of the Earth's energy budget that it cannot have significant effects.

3. In Paragraph 3, which of the sentences below best expresses the essential information in the highlighted sentence in the passage? (   )
   A. Almost all of the short-wavelength energy in the Earth's energy budget comes from solar radiation.
   B. Short-wavelength radiation is by far the largest part of the total energy that the Sun radiates to the Earth.
   C. The amount of short-wavelength radiation received from the Sun is huge by comparison to the Earth's own energy production.
   D. Almost the entire amount of energy that flows into the Earth's energy budget is short-wavelength radiation from the Sun.

4. According to Paragraph 4, all of the following statements about geothermal energy are true EXCEPT (   )
   A. it is the main source of heat for the surface of the Earth.
   B. it is responsible for earthquakes.
   C. it causes the eruptions of volcanoes.
   D. it causes mountains to rise high above the rest of the Earth's surface.

Unit 6　Resources and Energy on the Earth

5. The word "stationary" in Paragraph 5 is closet in meaning to　　　　　(　　)
   A. isolated.
   B. visible.
   C. raised.
   D. unmoving.
6. Paragraph 5 mentions which of the following as an effect of the Moon's gravitation on the Earth?　　　　　(　　)
   A. It causes high tides that reshape the coastlines of continents and islands.
   B. It causes the Earth to rotate on its axis at a somewhat faster speed than it would otherwise.
   C. It pulls ocean water into a bulge that runs into land masses as the Earth rotates on its axis.
   D. It reduces the force with which the tidal bulge would otherwise pile up against continents.
7. What can be inferred from Paragraph 6 about what would likely occur if cloud cover increased worldwide?　　　　　(　　)
   A. Different materials would become more similar to each other in their reflectivity.
   B. It would become a greater necessity to convert forests into plowed land and cities.
   C. A large percentage of incoming solar radiation would be reflected back into space.
   D. The reflectivity of ice and water would change and become greater over time.
8. According to Paragraph 7, weather patterns are produced as part of the cycle in which
   (　　)
   A. incoming solar energy becomes reflected back into space.
   B. solar energy is converted into geothermal and tidal energy.
   C. the atmosphere and hydrosphere absorb long-wavelength radiation.
   D. energy that has been absorbed near the Earth's surface undergoes transfer and conversion of form.
9. The letters [A], [B], [C], and [D] indicate where the following sentence (in bold) could be added to the following part of the passage. Where would the sentence best fit?　　(　　)

   > **How reflective a material is depends on how light or dark it is, among other things.**

   　　The Earth loses energy from the cycle in two main ways: reflection, and degradation and reradiation. About 40 percent of incoming solar radiation is simply reflected, unchanged, back into space by the clouds, the sea, and other surfaces. [A] For any planetary body, the percentage of incoming radiation that is reflected is called the "albedo". [B] Each different material has a characteristic reflectivity. [C] For example, ice is more reflectant than rocks or pavement; water is more highly reflectant than vegetation; and forested land reflects light differently than agricultural land. [D] Thus, if large expanses of land are converted from forest to plowed land, or from forest to city, the actual reflectivity of the Earth's surface, and hence its albedo, may be altered. Any change in albedo will, of course, have an effect on the

Earth's energy budget.

10. An introductory sentence (in bold) for a brief summary of the passage is provided below. Complete the summary by selecting the THREE answer choices that express the most important ideas in the passage.　　　　　　　　　　　　　　　　　　　　　(　　)

> **The Earth's energy cycle consists of all the energy inputs, outputs, and conversions within the Earth system, which must maintain an overall balance.**

A. Incoming short-wavelength solar radiation provides the Earth with nearly all its energy and powers the hydrologic cycle as well as biological processes.

B. Heat energy from the Earth's interior, which powers the rock cycle, and the kinetic energy of the Earth's rotation provide small additions to solar energy.

C. Some of the incoming solar radiation is reflected and the rest, after being absorbed, undergoes a series of conversions until it is reradiated into space as heat.

D. Humans use only a small amount of the available solar energy for heat, satisfying most of their energy needs by burning coal, oil, and natural gas.

E. Solar energy stored in rocks on the Earth's surface is the primary source of geothermal energy and tidal energy.

F. The Earth's atmosphere and hydrosphere absorb most of the incoming solar radiation, using up much of the energy to power weather patterns, with only the remainder radiated out as heat.

Passage C

Scan and read along the passage

Scan and read along the vocabulary

transformation *n.*（彻底或重大的）改观,变化,转变,核的转换
splitting *n.* 分裂
fission *n.* 裂变,分裂
bombard *n.* 轰炸,连环炮击,(物)以高速粒子轰击
neutron *n.* 中子

Nuclear Energy*

❶ Nuclear energy comes from the heat energy produced during the induced **transformation** of a chemical element into other chemical elements. In theory, nuclear energy can be generated in two ways: by inducing a heavy atom to split into lighter atoms, or by causing two light atoms to combine to make a heavier atom. In practice, only one of these (**splitting**) is usable for power generation with existing technology.

❷ Splitting heavy atoms into lighter atoms is called **fission**. Fission is induced by **bombarding** fissionable atoms with **neutrons**,

---

* This passage is adapted from SKINNER B J, MURCK B W, 2011. The blue planet: an introduction to Earth system science [M]. 3rd ed. Hoboken, NJ: John Wiley.

electronically neutral particles from the **nucleus** of an atom. When a fissionable atom is hit by a neutron, it splits into two different atoms, both of which are lighter than the original. Some matter is "lost" during the splitting process. However, because of the Law of Conservation of Matter and Energy$^{(1)}$, this matter is not actually lost but is instead transformed into energy. Thus, when the atom splits, it creates two new (lighter) atoms and releases the **leftover** energy as heat. The split atom also **ejects** some neutrons from its own nucleus. These neutrons can be used to induce more atoms to split, creating a chain reaction. The entire process is carried out inside a nuclear **reactor**. The rate of neutron bombardment is controlled, or **moderated**, usually by water. When a chain reaction proceeds without control, an atomic explosion occurs.

❸ ^{235}U is a naturally occurring fissionable material that is mined and used as fuel in nuclear reactors (in addition to other fissionable materials). The fissioning of just 1 gram of ^{235}U produces as much heat as the burning of 13.7 barrels of oil. The ^{235}U is processed and concentrated into fuel **pellets**, which are packed into a bundle of hollow tubes, called *fuel rods*. The fuel rods are loaded into the core of the reactor, where the fission process is induced. The heat generated by fission is carried away by water, which also moderates the chain reaction. The heated water makes steam, which turns a **turbine**, producing electricity. If the heat were not removed from the fuel bundle, it could get so hot that the reactor core would melt, releasing its radioactive contents; this is called a **meltdown**, and it is what happened at Chernobyl in Ukraine$^{(2)}$ (in the former Soviet Union) in 1986. Current nuclear reactor technologies are designed to minimize or eliminate the possibility of a meltdown.

❹ Nuclear power is considered to be a clean source of energy—even by some environmentalists—because it causes no harmful atmospheric emissions. Approximately 17 percent of the world's electricity is derived from nuclear power plants. In France, more than half of all electric power comes from nuclear plants; the proportion is rising sharply in some other European countries and in Japan. The reason for the increase is that Japan and most European countries do not have adequate supplies of fossil fuels to be **self-sufficient**. In North America, however, the growth of the nuclear industry appears to have **stalled**; this is partly a result of

nucleus n. 原子核,细胞核,核心,核
leftover n. 残存物,遗留物
   adj. 剩下的,多余的
eject v. 喷射,排出,弹出
reactor n. 反应器,(核)反应堆
moderate v. 缓和,使适中,使(中子)减速
pellet n. 小球
turbine n. 涡轮机,汽轮机
meltdown n. 核反应堆芯熔毁,(公司、机构或系统的)崩溃
self-sufficient adj. 自给自足的
stall v. 暂缓,搁置,停顿

intractable *adj.* 棘手的, 难处理的
by-product *n.* 副产品, 附带产生的结果
decommissioning *n.* 退役, 停运, 停止运作
radioactive decay 放射性衰变
toxic *adj.* 有毒的, 引起中毒的
persistent *adj.* 坚持不懈的, 持续的, 反复出现的
half-life *n.* 半衰期
fraction *n.* 分数, 小数, 小部分, 微量
dissipate *v.* (使某事物)消散, 消失, 挥霍, 耗费

negative public opinion stemming from nuclear catastrophes like Chernobyl and from the **intractable** problems associated with nuclear waste disposal.

❺ In normal operation, nuclear power plants generate very little environmental radiation; we are actually exposed to far more radiation per year from natural sources than from nuclear reactors. However, nuclear fission generates highly radioactive **by-products**—nuclear waste, which must be isolated from the biosphere and hydrosphere. This presents a technically difficult disposal problem that has not yet been resolved.

❻ Radioactive waste, or radwaste, is leftover radioactive materials or equipment from nuclear power plants, laboratories, and medical procedures, uranium mining, and the **decommissioning** of nuclear weapons. Radioactivity is the energy and energetic particles that are emitted during the natural transformation of one element into another element, called **radioactive decay**. Radioactivity can damage organisms, even causing death if the dose is high enough. It also causes genetic damage, that is, damage to the organism's offspring. Radioactive elements can be **toxic** in minute quantities, and they can be extremely **persistent**, remaining radioactive for many thousands of years in some cases. Furthermore, all organisms—including humans—lack any kind of natural warning for the presence of radioactivity; we can't smell it, see it, taste it, or feel it.

❼ The decay of radioactive materials is measured in units of **half-life**, the amount of time it takes for the level of radioactivity in a material to decrease by half (that is, to 50 percent of the original level). Thus, in two half-lives the radioactivity in the material will have decreased to 25 percent of its original level. In three half-lives, it will have decreased to 12.5 percent of its original level, and so on. Some radioactive materials have half-lives measured in seconds, or **fractions** of a second. Others have half-lives measured in days, years, thousands of years, even millions of years.

❽ Low-level radioactive wastes, which emit little radiation and have short half-lives, are contained and then released in a controlled manner once their radioactivity has **dissipated**. Much more problematic are the high-level radioactive wastes contained in spent fuel rods, which comprise only 1 percent of the waste generated by nuclear reactors, but generate 99 percent of the radioactivity. So far, there are no permanent **repositories** for high-level radioactive waste. Most high-level wastes are stored temporarily in concrete **bunkers** or water pools (because water dissipates the heat, as well as absorbing the radioactivity). There has been an intensive search, for the past two decades or more, for appropriate disposal sites and designs for permanent repositories for high-level nuclear waste.

❾ All permanent disposal options for high-level waste involve concentration and **immobilization** of the waste, usually in **inert** glass or **ceramic** pellets; containment in specially engineered **canisters**; and subsequent isolation from the biosphere and hydrosphere. An appropriate disposal site also must be isolated from potential resources and human activities, such as mining. It must be engineered so that it cannot be easily entered, damaged, or **sabotaged**. It must be economically feasible and capable of holding a large quantity of waste for a long time—10,000 years is commonly proposed. Ideally, disposal will not require transportation over long distances because the transport of **hazardous** materials is risky. Finally, the medium that surrounds the waste must be geologically stable and nonfractured (because fractures provide pathways along which escaping contaminants could travel), and must provide good heat **conductivity** (to dissipate the heat generated by the radioactive materials) and excellent chemical absorption.

❿ It's a tall order, but many scientists feel that it is technically possible to find an appropriate site and design a permanent disposal facility for nuclear waste that will meet these criteria. So far, all serious options for the permanent disposal of high-level nuclear waste involve land-based **geologic isolation**—that is, using the properties of natural rocks to isolate and contain the material. This is combined with a multiple barrier concept, in which the repository is engineered to place many physical and chemical barriers in the way of any escaping contaminants.

⓫ In principle, **nuclear fusion**—the joining together or fusing of two small atoms to create a single larger atom, with **attendant** release of heat energy—is another potential source of nuclear power. Nuclear fusion utilizes as its fuel a "heavy" isotope of hydrogen, called **deuterium**. The Earth has a virtually endless supply of deuterium in the form of a very common chemical

repository n. 贮藏室,仓库
bunker n. 沙坑,煤仓,燃料库
immobilization n. 使停止流通,固定化,使活动受限
inert adj. 惰性的
ceramic adj. 陶瓷的
canister n. 筒,小罐,防毒面具的滤毒罐
sabotage v. 蓄意破坏,故意毁坏,捣乱,阻挠
hazardous adj. 危险的,有害的
conductivity n. 传导性
geologic isolation 地质隔离
nuclear fusion 核聚变,核子融合
attendant adj. 伴随的,随之产生的
deuterium n. 氘,重氢

hydrogen dioxide 过氧化氢
helium n. 氦
tremendous adj. 巨大的,极大的,令人望而生畏的,可怕的
ambient adj. 环境的,周围的
grail n. 杯,圣杯,大盘,长期以来梦寐以求的东西

compound—water (**hydrogen dioxide**, H_2O_2). The primary byproduct of nuclear fusion would be **helium**, a non-toxic, chemically inert gas.

❷ This conjures up images of a cheap, clean, virtually inexhaustible power source. So, why are we not using energy provided by nuclear fusion? Fusion is the nuclear process that occurs in the cores of stars, the process responsible for the **tremendous** heat energy generated by the Sun. But that, in a nutshell, is the problem. For two atomic nuclei to fuse, the **ambient** conditions must be similar to those at the core of a star—on the order of 100 million degrees. The possibility that nuclei could be induced to fuse at something close to room temperature (so-called cold fusion) has led many scientists to search for this "holy **grail**" of energy, but routine use of fusion power remains an unrealized goal.

Notes

(1) **The Law of Conservation of Matter and Energy** are two laws in chemistry that are used to explain the properties of isolated, closed thermodynamic systems. These laws state that matter or energy cannot be created or destroyed but can be converted to different forms or get rearranged. The key difference between the two laws is that the law of conservation of matter states the total mass inside a closed system that does not allow matter or energy to escape should be a constant whereas the law of conservation of energy states energy cannot be created or destroyed, but may be changed from one form to another.

(2) **Chernobyl** is a nuclear power plant in Ukraine that was the site of a disastrous nuclear accident on April 26, 1986. A routine test at the power plant went horribly wrong, and two massive explosions blew the 1,000-ton roof off one of the plant's reactors, releasing 400 times more radiation than the atomic bomb dropped on Hiroshima.

Reading Comprehension

 Keys

Directions Fill in blanks to complete the summary of Passage C.

Nuclear energy can be generated by inducing a heavy 1._____ to split into lighter ones. 2._____ is carried out inside a nuclear reactor, using fissionable materials such as 3._____ in a chain reaction. Nuclear power is a clean source of energy, but it generates 4._____ that must be isolated from the 5._____ and 6._____ for thousands of years. Most proposals for the permanent disposal of 7._____ radioactive waste rely on

8._____. 9._____ is another a potential source of cheap, clean, virtually inexhaustible power, but 10._____ fusion remains an unrealized goal.

Exercises

Vocabulary

Ⅰ. Technical words

Directions Match the technical terms with their definitions.

A. renewable resources | a. resources that are found in a region, but have not been developed yet

B. biotic resources | b. form of energy conversion in which heat energy from within the Earth is captured and harnessed for cooking, bathing, space heating, electrical power generation, and other uses.

C. stock resources | c. short-wavelength solar radiation the Earth receives from the Sun

D. potential resources | d. energy comes from splitting atoms in a reactor to heat water into steam, turn a turbine, and generate electricity

E. solar energy | e. power produced by the surge of ocean waters during the rise and fall of tides

F. geothermal energy | f. resources that can naturally regenerate after use such as wind, water, plants, animals, and solar energy

G. tidal energy | g. living resources that exist naturally in the environment such as forests, wildlife, and fossil fuels

H. nuclear energy | h. materials in the environment which have the potential to satisfy human needs but human beings do not have the appropriate technology to access them

Ⅱ. Academic vocabulary

Directions Fill in the blanks with appropriate forms of the words given below. Each word can be used only once.

| harness | inorganic | depletion | minute | dissipate |
| survey | inert | extract | irreversible | degradation |

1. Any _____ in ozone levels is viewed with alarm because it would allow more dangerous radiation to reach the surface, increasing the risk of skin cancer, among other effects.
2. An ecosystem is a system involving interactions among living organisms and interactions between organisms and the _____ environment.
3. However, the Sun's energy is diffuse, and _____ it efficiently is a major challenge to widespread use of solar power.
4. Color is extremely variable in quartz and many other minerals because even _____ chemical impurities can strongly influence it.
5. Unlike gasoline, which is refined from petroleum, other fuels called synthetic fuels are _____ from solid organic material.
6. When drought occurs, as it inevitably does, and the vegetative cover has been destroyed beyond the minimum to hold the soil against erosion, the destruction becomes _____.
7. We see, therefore, a progression of activity from small, relatively _____ Mercury, to slightly larger and once-active the Mars, to the larger and far more active interiors and surfaces of the Venus and the Earth.
8. Both surface water and groundwater can suffer _____ and depletion as a result of withdrawals that exceed the rate of recharge.
9. Energy is not recycled in ecosystems because at each transformation much of it is _____ as heat, a form of energy that cannot be used by organisms to power their metabolism.
10. Long ago, geologists realized that anomalous gravity values should exist over buried bodies of ore minerals and salt domes and that geologic structures such as faulted strata could be located by surface gravity _____.

Sentence Structure

Understand participle clauses in English sentences.

Participle clauses are formed using present particles (going, reading, seeing, walking, etc.), past participles (gone, read, seen, walked, etc.) or perfect participles (having gone, having read, having seen, having walked, etc.).

Present participle clauses have a similar meaning to active verbs. They could be used when the participle and the verb in the main clause have the same subject.

For example,
  Waiting for Ellie, I made some tea. (While I was waiting for Ellie, I made some tea.)
  Knowing she loved reading, Richard bought her a book.

Past participle clauses usually have a passive meaning.

For example,
  Watered the right amount, plants can grow big and tall. (If plants are watered the right amount, they can grow big and tall.)
  Frightened by the noise, she turned on the light.

Unit 6 Resources and Energy on the Earth

Perfect participle clauses show that the action they describe was finished before the action in the main clause. Perfect participles can be structured to make an active or passive meaning.

For example,
  Having got dressed, he slowly went downstairs.
  Having finished their training, they will be fully qualified doctors.

It is also common for participle clauses, especially with *-ing*, to follow conjunctions and prepositions such as *before, after, instead of, on, since, when, while* and *in spite of*.

For example,
  Before cooking, you should wash your hands.
  Instead of complaining about it, they should try doing something positive.

Present participle phrases and gerund phrases are easy to confuse because they both begin with an *-ing* word. The difference is the function that they provide in a sentence. A present participle phrase will always act as an adjective while a gerund phrase will always behave as a noun.

For example,
  Walking on the beach, Delores dodged the jellyfish that had washed ashore.
  Walking on the beach is painful if jellyfish have washed ashore.

In the first sentence, *Walking on the beach* is a present participle phrase describing the noun Delores, while in the second sentence, *Walking on the beach* is a gerund phrase, working as the subject of the verb *is*.

Directions Analyze the following sentences by: 1) underlining the participle clause(s), and 2) identifying the type of participle clause used in each sentence.

1. Incoming short-wavelength solar radiation overwhelmingly dominates the flow of energy in the Earth's energy budget, accounting for about 99.986 percent of the total. (Paragraph 3, Passage B)
2. The force of the tidal bulge "piling up" against land masses acts as a very slow brake, actually causing the Earth's rate of rotation to decrease slightly. (Paragraph 5, Passage B)
3. Nuclear energy comes from the heat energy produced during the induced transformation of a chemical element into other chemical elements. (Paragraph 1, Passage C)
4. Splitting heavy atoms into lighter atoms is called fission. (Paragraph 2, Passage C)
5. If the heat were not removed from the fuel bundle, it could get so hot that the reactor core would melt, releasing its radioactive contents. (Paragraph 3, Passage C)
6. Radioactive elements can be toxic in minute quantities, and they can be extremely persistent, remaining radioactive for many thousands of years in some cases. (Paragraph 6, Passage C)

Translation

I. Translate the following English sentences into Chinese. Pay attention to the participle clause in each sentence.

1. Any depletion in ozone levels is viewed with alarm because it would allow more dangerous radiation to reach the surface, increasing the risk of skin cancer, among other effects.

2. An ecosystem is a system involving interactions among living organisms and interactions between organisms and the inorganic environment.

3. However, the Sun's energy is diffuse, and harnessing it efficiently is a major challenge to widespread use of solar power.

4. Unlike gasoline, which is refined from petroleum, other fuels called synthetic fuels are extracted from solid organic material.

II. Translate the following Chinese paragraph into English.

我们星球的能量源于地球内部和外部。内部能量来源来自放射性同位素的地热能与地球旋转产生的旋转能；而太阳则是最主要的外部能源，影响了天气和气候等系统。阳光不同程度地加热地球表面与大气层，引起了对流，带来了风并影响了海洋气流。从地球温暖的表面散发出来的红外线辐射，被温室气体捕获住，从而进一步影响了能量流动。

Unit 7 Low Carbon Economics

A low-carbon economy or decarbonized economy is an economy based on energy sources that produce low levels of greenhouse gas (GHG) emissions. GHG emissions due to anthropogenic (human) activity are the dominant cause of observed climate change since the mid-20th century. In this unit, you will read about: 1) a brief history of energy industry, 2) smart energy technology for harvesting renewable energy, and 3) what geoscientists can do for decarbonization.

Passage A

Scan and read along the passage

Energy, Society and Environment[*]

❶ In ancient times and right up to the beginning of the industrial revolution, motive power was provided either by direct human effort or by animals or, at sea, by wind. Wood was used for heating, cooking, and some processing of materials. In the Middle Ages, machines such as windmills and watermills gradually took on some of the load, and watermills actually played a key role in the early stages of the industrial revolution. However, the industrial revolution only really got going when coal began to be used widely. Whereas, before, factories had to be sited near power sources such as rivers and also possibly near sources of raw materials, coal could be transported to factories—in trains powered by coal-fueled steam engines—to run the factory equipment. Coal also provided a fuel for trains and ships to shift raw materials from remote sites to factories and to transport their finished products to distant markets. Moreover, coal-fired engines

Scan and read along the vocabulary

---

[*] This passage is adapted from ELLIOT D, 2003. Energy, society and environment[M]. London: Routledge.

breakthrough *n.* 突破,重大进展
combustion *n.* 燃烧,氧化
overtake *v.* 追上,超过
hydroelectric *adj.* 水力发电的
concentrate *v.* 集中注意力,聚集,浓缩
carbon *n.* 碳
dioxide *n.* 二氧化物

could be used to pump water out of mines, thus allowing deeper mines to be used and more coal to be produced. The combination of coal, steam engine, and train technology was thus a real industrial and economic **breakthrough**.

❷ The role of coal as the key fuel was not challenged for hundreds of years until the beginning of the 20th century when the petrol-powered internal **combustion** engine had been invented and widely used. In the inter-war years, while electricity became increasingly important in many sectors of the economy, oil and gas also gradually took on increasing roles. After the Second World War, oil began a steady rise and, by the time of the first oil crisis in 1973—1974, its use had **overtaken** coal in most advanced industrialized countries. As the modern industrialization process got underway, the basic energy pattern was fairly stable: coal, oil, and gas had very roughly equal shares in primary energy use terms, but with oil beginning to dominate and with nuclear power, along with **hydroelectric** plants, making a small additional contribution. Basically, coal provided the bulk of electricity, gas provided the bulk of heating, oil the bulk of transport fuel. The exact proportions differed from country to country, depending on the availability of indigenous reserves. For example, the UK (the United Kingdom) had ample coal, while the USA (the United States of America) had coal, oil, and gas. By contrast, some of the newly emerging countries like Japan had few coal reserves and had to rely heavily on imported oil.

❸ Ever since mankind started burning wood, environmental impact has been a problem, but it became worse as the population increased and was more **concentrated** as industrialization expanded. For example, air pollution from coal burning had emerged as a key issue in the UK and European mainland as a result of a series of disastrous "smog" in the 1950s. Although coal has been burnt in a cleaner way thanks to technology development in the recent decades, another issue emerged as climate change that inflicted much more concern on the use of fossil fuels. All fossil fuels create **carbon dioxide** (CO_2) when burnt. Greenhouse gases like CO_2 travel up into the troposphere where they act as a screen to sunlight. They allow the sun's rays in but stop the heat radiation from re-emerging, resulting in what is called the greenhouse effect. Some degree of greenhouse effect is vital for the planet to support life. However, the vast **tonnage** of carbon dioxide gas we have released into the atmosphere seems likely to upset the natural balance and give rise to excessive global warming. It has been observed that the CO_2 level in the upper atmosphere reached 413 parts per million (ppm) in 2020 (IEA, 2020). This is 47% more than that in the pre-industrial era around 1850 and, if the trend continues, will possibly reach 1,000 ppm by the end of this century.

❹ The situation has been made worse by the fact that "large" areas of forest around the world have been felled, thus weakening an important "sink" for carbon dioxide. Re-**afforestation** is obviously vital, but it would take many decades just to replace what has been lost. What's more, unless carried out on a vast scale, the creation of new forest could only play a small role in absorbing the vast amount of carbon dioxide released every year by power plants and cars.

❺ The results of global warming, if it happens on a significant scale, are likely to be highly severe. One result is the ice-caps melting that gives rise to significant sea-level rises. Global warming could also lead to extreme weather such as storms, **droughts**, and flooding, which could cause great loss of life and economics. The last decade has seen some dramatic changes in weather and climate patterns, and now the scientific **consensus** is that this is the result primarily of human activity.

❻ There has however been some **enthusiasm** for a return to the idea of using nuclear power. But the scale of nuclear expansion would have to be very large in order to make a significant impact. It seems unlikely since in most countries the low-carbon benefit has hardly **outweighed** the public concern on security and problems in managing the radioactive wastes[1]. Following the Chernobyl accident in Ukraine in April 1986, support for nuclear power worldwide dropped notably. For example, a National Opinion poll in UK carried out in 2002 found that 85 percent wanted government investment in "eco-friendly" renewable energy (solar, wind, and water power) and only 10 percent support governmental investment in building new nuclear plants. It does seem that nuclear power is not popular with most people. Public opposition on this scale has clearly affected decision making about nuclear power, with governments being aware of its unpopularity.

tonnage *n.* 吨位,总质量,总吨数
afforestation *n.* 造林
droughts *n.* 干旱
consensus *n.* 一致看法,共识
enthusiasm *n.* 热情,热忱
outweigh *v.* 超过,比……重

environmentalist *n.* 环境保护主义者,环境保护论者,环境艺术家

undermine *v.* 逐渐削弱,故意破坏……的形象,在……下面挖,从根基处损坏

diffuse *adj.* 扩散的,弥漫的,难解的,冗长的

intermittent *adj.* 间歇的,断断续续的

photovoltaic *adj.* 光伏的,光电的

cladding *n.* 包层,外墙

maintenance *n.* 维护,保养,维持

Environmentalist pressure on safety issues forced up the price of nuclear power, and that, along with the relatively lower price of fossil fuel, further **undermined** the economics of the industry.

❼ Rather than trying to create little artificial suns on the Earth in the form of fusion reactors, research on solar power and other forms of renewable energy has expanded and has led to increasingly largescale deployments. Renewable energy, unlike fossil fuels, are naturally replenished. Although there can be problems with trying to use what are generally more **diffuse** and **intermittent** energy sources, new technologies for harvesting renewable energy are developing rapidly, including solar power and wind power, etc.

❽ Solar energy can be harvested in various ways, such as solar heat-concentrating mirrors and roof-top solar collectors. However, the big breakthrough is likely to be in the **photovoltaic** (PV) solar field. Photo cells, like those on cameras and pocket calculators, convert sunlight directly into electricity. As a convenient substitute for roof top or building **cladding** to save cost, PV cells were then found especially useful for distributed power generation in areas where large power grid is not readily accessible. In recent years the costs are dropping rapidly as new types of cell materials with higher efficiency have been developed. Commercial modules can be up to 25 percent efficient[2], and some laboratory devices have efficiencies of more than 47 percent.

❾ Due to economic and environmental benefits, PV cells have been widely used in many countries. In Germany, about 1.9 million photovoltaic systems were installed in 2020, accounting for an estimated 8.5 percent of the country's gross-electricity generation. It is increasingly seen as being likely to be a major energy source in the decades ahead.

❿ In contrast to PV, wind power is already a significant energy source. By the year 2020, there have been 743 gigawatt (GW) of wind power capacity installed worldwide, accounting for about 11 percent of global electricity generation capacity as well as a contribution of avoiding over 1.1 billion tons of CO_2 globally.

⓫ Wind turbines are often grouped together in "wind farms", so that connections to the power grid can be shared, as well as control systems and road access for **maintenance**. Typically, a separation distance between 5 and 15 times blade diameters is needed between individual wind turbines to prevent **turbulent** interactions in wind farm arrays. This means that wind farms can take up quite a lot of space, leading to some objections. It is argued that there would be insufficient room in countries like the UK to generate significant amounts of power. Besides, since wind turbines are visually **intrusive**, there will be specific siting constraints which could reduce net power availability. Compared with onshore wind power, the offshore potential is much less constrained in spite of its unpredictability. If the 2050 target of the International Renewable Energy Association of 2 terawatt (TW) of offshore wind in a net-zero world is to be met, it is predicted that Asia will be the home to nearly 40% of installations, followed by Europe on 32%.

❶❷ Although the sun delivers a vast amount of energy to the Earth, what could ever be used is much less. What matters is the extent to which renewable energy technology can be developed in practical terms and the timescale on which this might be achieved. That of course will depend on, amongst other things, the success of the technological development program, the economics, and the environmental constraints, and, crucially, for a new area of technology, on the level of support given.

turbulent *adj.* 骚乱的,湍流的

intrusive *adj.* 侵入的,打扰的

Notes

(1) **Radioactive Waste Management:** In nuclear plants about 96% of spent nuclear fuel is recycled back into uranium-based and mixed-oxide fuels. The residual 4% is fission products which are highly radioactive High-Level Waste. This radioactivity naturally decreases over time, so the material is stored in appropriate disposal facilities for a sufficient period until it no longer poses a threat.

(2) **Solar cell efficiency** refers to the portion of energy in the form of sunlight that can be converted via photovoltaics into electricity by the solar cell. In 2019, the world record for solar cell efficiency at 47.1% was achieved by using multi-junction concentrator solar cells, developed at National Renewable Energy Laboratory, Golden, Colorado, USA.

Reading Comprehension

Directions Answer the following questions according to Passage A.

1. According to Paragraph 1, what played a key role in the early stage of the industrial revolution? ( )
   A. Steam engines.
   B. Trains.
   C. Watermills.
   C. Coal.

2. The word "indigenous" in Paragraph 2 is closest in meaning to ( )
   A. native.
   B. clever.
   C. personal.
   D. overall.

3. Which of the following regarding oil is not mentioned in Paragraph 2? ( )
   A. Use of oil began a steady rise after the Second World War.
   B. Oil crisis took place in October 1973.
   C. Oil began to dominate during the industrialization process.
   D. Oil contributed to the bulk of energy consumption in transportation.

4. Which of the following cannot be inferred according to Paragraph 3? ( )
   A. A mild degree of greenhouse effect is important to the lives on the Earth.
   B. Greenhouse gas include carbon dioxide, water vapor, methane, etc.
   C. Hazardous pollutants in burning coal have been reduced.
   D. Carbon emission has greatly increased due to human activities.

5. What does the author think of global warming? ( )
   A. Forests act as carbon sink to aggravate global warming.
   B. Re-afforestation does not help to control global warming.
   C. The result of global warming is serverer at any rate.
   D. Significant sea-level rise is caused by ice-caps melting.

6. In Paragraph 6, the author mentioned the Chernobyl accident in order to ( )
   A. advocate the using of nuclear power.
   B. explain why support for nuclear power decreased.
   C. agitate public concern on the security of nuclear power.
   D. call for the development of other renewable energy.

7. Which of the following is not the reason why nuclear power didn't get significant expansion in most countries according to Paragraph 6? ( )
   A. Public concern on security.

B. Problem in managing radioactive wastes.

C. Price disadvantage.

D. Prohibition from policy-makers.

8. Which of the following is not the reason why wind farms are objected?　　(　　)

   A. It is difficult to forecast wind energy.

   B. They may occupy too much lands.

   C. They can cause disruption or annoyance in sight.

   D. Onshore wind power potential is constrained.

Passage B

 Scan and read along the passage

Smart Energy*

 *Scan and read along the vocabulary*

❶ The next few decades will see great changes in the way energy is supplied and used. In some major oil producing nations, "peak oil" has already been reached, and there are increasing fears of global warming. Consequently, many countries are focusing on the switch to a low carbon economy. This **transition** will lead to major changes in the supply and use of electricity. Firstly, there will be an increase in overall demand, as consumers switch from oil and gas to electricity to power their homes and vehicles. Secondly, there will be an increase in power generation, not only in terms of how much is generated, but also how it is generated, as there is growing electricity generation from renewable sources. To meet these challenges, countries are investing in Smart Grid technology[1]. This system aims to provide the electricity industry with a better understanding of power generation and demand and to use this information to create a more efficient power network.

transition *n.* 过渡, 转变

❷ Smart Grid technology basically involves the application of a computer system to the electricity network. The computer system can be used to collect information about supply and demand and improve engineer's ability to manage the system. With better information about electricity demand, the network will be able to increase the amount of electricity delivered per unit generated, leading to potential reductions in fuel needs and carbon emissions.

---

* This passage is adapted from ISMANTHONO H W, 2015. TOEFL for your careers[M]. Jakarta: Gramedia Widya Sarana Indonesia.

appliance *n.* 家用电器,装置
incentive *n.* 激励,刺激
notoriously *adv.* 声名狼藉地,臭名昭著地

Moreover, the computer system will assist in reducing operational and maintenance costs.

❸ Smart Grid technology offers benefits to the consumer too. They will be able to collect real-time information on their energy use for each **appliance**. Varying tariffs throughout the day will give customers the **incentive** to use appliances at times when supply greatly exceeds demand, leading to great reductions in bills. For example, they may use their washing machines at night. Smart meters can also be connected to the internet or telephone system, allowing customers to switch appliances on or off remotely. Furthermore, if houses are fitted with the apparatus to generate their own power, appliances can be set to run directly from the on-site power source, and any excess can be sold to the grid.

❹ With these changes comes a range of challenges. The first involves managing the supply and demand. Sources of renewable energy, such as wind, wave, and solar, are **notoriously** unpredictable, and nuclear power, which is also set to increase as nations switch to alternative energy sources, is inflexible. With oil and gas, it is relatively simple to increase the supply of energy to match the increasing demand during peak times of the day or year. With alternative sources, this is far more difficult, and may lead to blackouts or system collapse. Potential solutions include investigating new and efficient ways to store energy and encouraging consumers to use electricity at off-peak times.

❺ A second problem is the fact that many renewable power generation sources are located in remote areas, such as windy uplands and coastal regions, where there is currently a lack of electrical infrastructure. New infrastructures therefore must be built. Thankfully, with improved smart technology, this can be done more efficiently by reducing the reinforcement or construction costs.

❻ Although Smart Technology is still in its infancy, pilot schemes to promote and test it are already underway. Consumers are currently testing the new smart meters which can be used in their homes to manage electricity use. There are also a number of demonstrations being planned to show how the smart technology could practically work, and trials are in place to test the new electrical infrastructure. It is likely that technology will be added in "layers", starting with "quick win" methods which will provide initial carbon savings, to be followed by more advanced systems at a later date. Cities are prime candidates for investment into smart energy, due to the high population density and high energy use. It is here where Smart Technology is likely to be promoted first, utilizing a range of sustainable power sources, transport solutions and an infrastructure for charging electrically powered vehicles. The infrastructure is already changing fast. By the year 2050[2], changes in the energy supply will have transformed our homes, our roads and our behavior.

Notes

(1) **Smart Grid technology** is the digital technology that allows for two-way communication between the utility and its customers, and the sensing along the transmission lines is what makes the grid smart.

(2) **2050** is the year by which many countries have promised to become carbon neutral.

Reading Comprehension

  Keys

Directions   Answer the following questions according to Passage B.

1. According to Paragraph 1, what has happened in some oil producing countries?   ( )
   A. They are unwilling to sell their oil any more.
   B. They are not producing as much oil as they used to.
   C. The supply of oil is unpredictable.
   D. Global warming is more severe here than in other countries.

2. In Paragraph 3, the author mentioned using washing machine at night in order to   ( )
   A. describe how smart grid works for various home appliances.
   B. encourage the reader to wash clothes at night to save money.
   C. prove the ability of smart grid to operate on intermittent energy sources.
   D. explain how smart grid helps people to reduce electricity bills.

3. Which of the following is NOT a benefit of Smart Grid technology to consumers?   ( )
   A. It can reduce their electricity bills.
   B. It can tell them how much energy each appliance is using.
   C. It can allow them to turn appliances on and off when they are not at home.
   D. It can reduce the amount of energy needed to power appliances.

4. According to Paragraph 4, what is the problem with using renewable sources of power?
   ( )
   A. They do not provide much energy.

B. They often cause system failure and blackouts.

C. They do not supply a continuous flow of energy.

D. They can't be used at off-peak times.

5. The word "remote" in Paragraph 5 could be best replace by ( )

   A. isolated.

   B. crowded.

   C. attractive.

   D. alone.

6. According to Paragraph 6, what of the following can be inferred about cities in the future?

   ( )

   A. More people will be living in cities in the future than nowadays.

   B. People in cities will be using cars and buses powered by electricity.

   C. All buildings will generate their own electricity.

   D. Smart Grid technology will only be available in cities.

7. The word "underway" in Paragraph 6 is closest in meaning to ( )

   A. permanent.

   B. complete.

   C. beneficial.

   D. in progress.

8. What is the main idea of Paragraph 6? ( )

   A. To describe who will benefit from Smart Grid technology first.

   B. To outline the advantages of Smart Grid technology.

   C. To summarize the main ideas in the previous paragraphs.

   D. To describe how, where and when Smart Technology will be introduced.

9. The letters [A], [B], [C], and [D] indicate where the following sentence (in bold) could be added to the following part of the passage. Where would the sentence best fit?

   > **There is also likely more electricity generation centers, as households and communities take up the opportunity to install photovoltaic cells and small-scale wind turbines.**

   The next few decades will see great changes in the way energy is supplied and used. In some major oil producing nations, "peak oil" has already been reached, and there are increasing fears of global warming. Consequently, many countries are focusing on the switch to a low carbon economy. This transition will lead to major changes in the supply and use of electricity. [A] Firstly, there will be an increase in overall demand, as consumers switch from oil and gas to electricity to power their homes and vehicles. [B] Secondly, there will be an increase in power generation, not only in terms of how much is generated, but also how it is generated, as there is growing electricity generation from renewable sources. [C] To meet these challenges, countries are investing in Smart Grid technology. [D] This system aims to provide the electricity industry with a better understanding of power generation and demand, and to use this information to create a more efficient power network.

10. An introductory sentence (in bold) for a brief summary of the passage is provided below. Complete the summary by selecting the THREE answer choices that express the most important ideas in the passage.

> **Many countries are investing in Smart Grid technology which helps in creating a more efficient electricity network.**

A. Smart Grid technology is an important helper for consumers to save electricity bills.

B. Pilot schemes are underway to test the new electrical infrastructure in our homes and cities.

C. While we must deal with the unpredictability and additional construction costs of renewable energy, new technologies will bring dramatic changes to our lives.

D. Smart Grid technology enables the consumers to better understand power generation and demand.

E. The inflexible energy sources and the construction of new infrastructures lead to a range of challenges.

F. For electricity industry, the application of computer system provides valuable information about electricity demand.

Passage C

Scan and read along the passage

Geoscience and Decarbonization: Current Status and Future Directions[*]

Scan and read along the vocabulary

❶ Geoscience has long been understood as part of the solution to decarbonization. A paper in *Science* magazine "Stabilization **wedges**: Solving the climate problem for the next 50 years with current technologies" by Pacala & Socolow (2004) established the important concept that a number of **complementary** technological fixes and behavioral changes could be used to bring about emissions reduction of a size that can make a difference for climate change. Their concept visualized CO_2 emissions reduction as a set of activities that have a geoscience aspect such as geological

wedge *n.* 楔形物,三角形物

complementary *adj.* 互补的,相辅相成的

---

[*] This passage is adapted from STEPHENSON M H, STEPHENSON M H, RINGROSE P, et al., 2019. Geoscience and decarbonization: current status and future directions[J]. Petroleum Geoscience, 25: 501-508.

subsurface *adj.* 地下的，表面下的
borehole *n.* 钻孔，井眼
enthalpy *n.* 焓，热函
magma *n.* 岩浆
hydrothermal *adj.* 热液的，热水的
cavern *n.* 洞穴，凹处

controls on nuclear waste disposal in increased nuclear scenarios, increased supply of gas to allow a switch of power generation from coal to gas in thermal power stations and carbon capture and storage (CCS). It was within this background of the urgent need for global emissions reductions that 100 geoscientists, social scientists, and policy-makers met at the 2019 Bryan Lovell meeting to offer geological solutions to the "well below 2 ℃" objective agreed at the COP21[1] conference in Paris. The main scientific themes included thermal storage, compressed air energy storage, geothermal energy, etc. More general themes at the conference included common challenges to geoscientists and the availability of geological skills for the geoscience decarbonization future.

❷ Thermal storage is an important technology for geoscience, decarbonization being critical for the energy used in domestic heating and cooling. Sebastian Bauer and Andreas Dahmke, from Christian-Albrechts University in Kiel, described the possibility of seasonal storage of large amounts of heat from solar or industry. Technical options for **subsurface** heat storage include aquifer and **borehole** thermal energy storage, which in principle enable heat storage in most subsurface geological formations. Using temperatures of up to 90 ℃ allows an increase in storage rates and capacities. To enable the implementation of large-scale urban subsurface heat storage, however, methods for dimensioning the storage systems in terms of achievable heat injection and extraction rates, as well as storage capacities, are required. Thomas Driesner (ETH Zurich) was looking for high **enthalpy** heat and described the potential of "superhot" geothermal in Iceland at a depth of 2 km immediately above a **magma** body, producing superheated steam reaching 450 ℃ and 140 bar at the wellhead. At the Larderello Field in Tuscany, described by Adele Manzella of CNR Italy, two European projects are also looking at deep chemical-physical conditions in an area characterized by very high heat flow in one of the most productive **hydrothermal** systems in the world.

❸ The challenge of intermittency raises the problem of energy storage, for which compressed air energy storage (CAES) can be a solution. In CAES, the idea is to store large amounts of compressed air in underground **caverns**—mainly in salt layers—for extraction through a turbine later. It has advantages over grid-scale batteries including longer lifetimes of pressure vessels and lower material **toxicity**. However, cavern design and construction are expensive. Another challenge is that air heats up when compressed from atmospheric pressure, and in an industrial CAES situation a storage pressure of about 70 bar is envisaged. Heat must be controlled to avoid damage to compressors and caverns. Salt caverns are favored because, being **impermeable**, there are no pressure losses, and because there is no reaction between the oxygen in the air and the salt. CAES is also feasible in natural aquifers, although oxygen may react with minerals in the host rock, and microorganisms in an aquifer can deplete oxygen and alter the character of the stored air; similarly, bacteria can act to block pore spaces in the reservoir. **Depleted** natural gas fields could also be used for CAES, although any mixing of residual hydrocarbons with compressed air would have to be considered.

❹ Pumped hydroelectric schemes (PHSs) are very suitable for rapid response, grid-scale energy storage. Martin Smith of the British Geological Survey explained that in the UK there are four such schemes located in Scotland and North Wales providing a maximum power output of 2.8 GW to the UK electrical grid. The main challenges for any PHS site include the topography, water availability and geology. Located in areas of predominantly ancient hard crystalline basement or volcanic rock, the geology is often assumed to be stable and predictable, but this is not always the case, necessitating in-depth geotechnical feasibility and mitigation studies. Smith described the detailed fault studies that were required following the failure of a tunnel due to **fracturing** and faulting, following stress release, at the Glendoe Hydroelectric Scheme. This resulted in the requirement of the construction of a bypass tunnel and a lengthy court action on the liability for the costs.

❺ Nuclear power is an energy source that has a strong, although indirect, connection with geoscience because waste produced will probably have to be **disposed** of in secure, deep geological repositories. Nuclear is widely considered to be a contributor to low-carbon power production, but also generate radioactive waste. Essentially, a geological disposal facility (GDF) makes use of engineered materials and structures including concrete, metals and clays, as well as the surrounding geological environment, as

toxicity *n.* 毒性
impermeable *adj.* 不可渗透的,透不过的
depleted *adj.* 不足的,耗尽的
fracturing *n.* 水力压裂,破碎,龟裂
dispose *v.* 处理,放置,安排

glaciation *n.* 冰川作用
notwithstanding *prep.* 虽然，尽管

containment barriers. Geoscientific expertise will play a central role in designing and siting the GDF and modeling the near-field response of the geosphere. A fundamental requirement of the geological environment is that its behavior should be predictable enough to establish very long-term radiological safety. Amongst the factors that need to be assessed are present and future levels of seismic activity, effects of **glaciation**, uplift and erosion, and future effects of climate change including sea-level rise because all of these processes could compromise the GDF.

❻ One aim of the conference was to understand some of the geological and scientific questions that are common to geological decarbonization technologies. One of the most fundamental challenges is the need to characterize rock geo-chemically and geo-mechanically. Rock formations, from unconsolidated sediment to sedimentary, metamorphic, and igneous rocks, needs to be understood to predict the performance of these materials in hosting energy-related systems such as low-enthalpy geothermal reservoirs or "hot-dry-rock reservoirs", the storage of CO_2 and other gases, and tunneling for pumped storage construction. A second common challenge recognized by scientists at the Bryan Lovell conference concerned the need to understand better the flow of fluids in the deep subsurface, whether they be water, carbon dioxide, natural gas or hydrogen. This is not a trivial task given the presence in the subsurface of several fluid phases, reactive rock, fractures and rock heterogeneity. Flow is important because in technologies like geothermal we want to encourage the flow of useful fluids (hot water), while in other technologies we want to contain fluids, such as in carbon dioxide capture and storage. An ability to monitor and verify through sophisticated imaging and detection will also be needed.

❼ **Notwithstanding** these challenges, the conference attendees agreed that the UK and Europe are very well placed to develop subsurface decarbonization technologies. The UK has excellent subsurface capability in its research base of world-class universities, research institutes, and oil and gas companies, and is also developing its experimental and pilot-scale infrastructure. Germany has significantly stepped up its efforts in renewable energy, with a growing geoscience focus on seasonal storage (hydrogen, air), while Norway has often taken the lead in developing CCS technology, operating the world's largest CO_2 capture test center (TCM Mongstad). EU research funding commitments in clean energy are very substantial, **spearheaded** by the Horizon Europe[2] program. These geoscience research and technology developments will certainly be required to enable us to decarbonize the present world energy system in tandem with surface renewables such as solar and wind (Stephenson, 2018). Indeed, the apparently "hidden subsurface" may offer the only solution to hard-to-decarbonize parts of the system including heavy industry such as steel, cement, and refineries, via CO_2 capture and storage technologies.

spearhead v. 带头，做先锋

❽ Critical to the success of the decarbonization initiative is knowledge and data-sharing across geographical borders, between industries and by all stakeholders of the subsurface—ensuring that competing interests are well managed. A successful and innovative set of subsurface decarbonization technologies developed in Europe will be an exportable asset in the years to come, leading to jobs, investment, and economic growth, and we expect to see geoscience playing a vital role.

Notes

(1) **The 21st yearly session of the Conference of the Parties** (COP 21) negotiated the Paris Agreement that targeted to keep the rise in mean global temperature to well below 2 ℃ above pre-industrial levels and preferably limit the increase to 1.5 ℃, recognizing that this would substantially reduce the effects of climate change.

(2) **Horizon Europe** is a 7-year European Union scientific research initiative, a successor of the recent Horizon 2020 program and the earlier Framework Programs for Research and Technological Development. The European Commission drafted and approved a plan for Horizon Europe to raise EU science spending levels by 50% over the years 2021–2027.

Reading Comprehension

Keys

Directions Match the paragraphs with their appropriate subtitles.

1. Paragraph 1 _____   A. Geological challenges for pumped hydroelectric schemes

2. Paragraph 2 _____   B. Two common challenges for geological decarbonization technology

3. Paragraph 3 _____   C. Heat storage technology for geoscience

4. Paragraph 4 _____   D. The role geoscientists should play in managing nuclear waste

5. Paragraph 5 _____   E. The background and main themes of the conference

6. Paragraph 6 _____   F. Importance of geoscience in the future of decarbonization

7. Paragraph 7 _____   G. Compressed air energy storage and its geoscientific considerations

8. Paragraph 8 _____   H. Opportunities and development of geosciences in UK and EU

Exercises

Vocabulary

Ⅰ. Technical words

Directions Translate the English expressions into appropriate Chinese versions.

sediment   _____
radioactive   _____
magma   _____
troposphere   _____
aquifer   _____
seismic   _____
cavern   _____
heterogeneity _____

Ⅱ. Academic vocabulary

Directions Fill in the blanks with appropriate forms of the words given below. Each word can be used only once.

| dispose | replenish | constrain | undermine |
| envisage | breakthrough | concentrate | consensus |

1. Six scientific _____ had changed the nature of almost all Yilgarn gold exploration thinking by 1990 and appear to have been important components of the exploration success.

2. The study of mineralogical variations is important and allows us to better understand the difficulties in reaching the chemical specifications of heavy mineral _____.

3. The American Geophysical Union (AGU) issues position statements reflecting the state of the science and scientific _____.

4. Offshore sediment supply to _____ beaches is transported nearshore and is often deposited as sediment berms and sand bars, the formation and stability of which can be impacted by storm events.

5. The solid cuttings are then either treated or _____ of by grinding or injecting them into slurries or keeping them in a waste pit for further treatment and disposal.
6. These observations indicate that year-round bottom-intensified inflow of warm Atlantic water through a narrow channel is _____ by a sill.
7. Tunneling in a karst environment can severely deplete an aquifer and _____ the sustainability of water resources over the long term.
8. The evolution of geology as an independent science can be _____ as a relatively continuous process yet marked by three fundamental steps.

Sentence Structure

Understand attributive clauses in English sentences.

An attributive clause modifies a word, phrase, or idea in the main clause. The word, phrase, or idea modified is called the **antecedent** (先行词). The most common relative pronouns are **who/whom, whoever/whomever, whose, that**, and **which**. Generally, there are two types of attributive clauses: **restrictive (defining) clause** (限定性定语从句) and **non-restrictive (non-defining) clause** (非限定性定语从句). In both types of clauses, the relative pronoun can function as a subject, an object, or a possessive pronoun (**whose**).

(1) **Relative Pronouns in Restrictive Attributive Clauses.** Relative pronouns that introduce a restrictive attributive clause ARE NOT separated from the main clause by a comma. Restrictive attributive clauses (also known as defining attributive clauses) add essential information about the antecedent in the main clause. The information is crucial for understanding the sentence's meaning correctly and cannot be omitted.

Relative pronouns used as a subject of a restrictive attributive clause:

> For example,
> This is the house *that had a great Christmas decoration*. (The relative pronoun "that" modifies the antecedent "house".)

Relative pronouns used as an object in a restrictive attributive clause:

> For example,
> This is the man to *whom I wanted to speak and whose name I had forgotten*. (The relative pronoun "whom" modifies the antecedent "man".)

(2) **Relative Pronouns in Non-Restrictive Attributive Clauses.** Relative pronouns that introduce non-restrictive attributive clauses ARE separated from the main clause by a comma (in most instances). Typically, "which" is the preferred relative pronoun for indicating that an attributive clause is non-restrictive. Non-restrictive attributive clauses (also known as non-defining attributive clauses) provide non-essential information about the antecedent in the main clause.

Relative pronouns used as a subject of a non-restrictive attributive clause:

> For example,
>
> The science fair, *which lasted all day*, ended with an awards ceremony. (The relative pronoun "which" modifies the antecedent "science fair".)

Relative pronouns used as an object in a non-restrictive attributive clause:

> For example,
>
> The sculpture, *which he admired*, was moved into the basement of the museum to make room for a new exhibit. (The relative pronoun "which" modifies the antecedent "sculpture".)

Directions   Analyze the following sentences by: 1) underlining the attributive clauses, 2) finding out the relative pronouns and the antecedents, and 3) figuring out whether each clause is a restrictive attributive clause or non-restrictive attributive.

1. One aim of the conference was to understand some of the geological and scientific questions that are common to geological decarbonization technologies. (Paragraph 6, Passage C)

2. Sources of renewable energy, such as wind, wave, and solar, are notoriously unpredictable and nuclear power, which is also set to increase as nations switch to alternative energy sources, is inflexible. (Paragraph 4, Passage B)

3. What matters is the extent to which renewable energy technology can be developed in practical terms, and the timescale on which this might be achieved. (Paragraph 13, Passage A)

Translation

I. Translate the following English sentences into Chinese. Pay attention to the attributive clause in each sentence.

1. One aim of the conference was to understand some of the geological and scientific questions that are common to geological decarbonization technologies.

2. What matters is the extent to which renewable energy technology can be developed in practical terms and the timescale on which this might be achieved.

3. Technical options for subsurface heat storage include aquifer and borehole thermal energy

storage, which in principle enable heat storage in most subsurface geological formations. (Paragraph 2, Passage C).

4. Although coal has been burnt in a cleaner way thanks to technology development in the recent decades, another issue emerged as climate change that inflicted much more concern on the use of fossil fuels. (Paragraph 3, Passage A).

II. Translate the following Chinese paragraph into English.

CO_2 地质封存是全球负排放的重要选项之一。海洋地质碳封存是把大规模排放源捕集而来的 CO_2 通过管线和井筒注入到海床表面或海洋地层里面从而实现 CO_2 长期隔离的一种气候工程前沿技术。国际上与碳捕集、利用与封存相关的海洋地质技术主要分为两大类:底地层 CO_2 单纯封存技术,以及利用 CO_2 来强化海洋油气资源的开采从而实现动态封存的 CO_2-EOR 技术。

Unit 8　Natural Disasters

The growing presence of natural disasters, a research domain of Geology, has provoked increasing attention. In this unit, you will read about different types of natural disasters from the following perspectives: 1) floods estimation and controlling system, 2) volcano monitoring, and 3) mitigating impacts on the Earth by near-Earth objects (NEOs).

Passage A

encroach (on) v. 侵蚀,蚕食(土地)
hazard n. 危险,危害
monitor v. 监控,监视,监听,检查
discharge n. 排出,流出
hydrologist n. 水文专家,水文工作者
designate v. 命名,标示
recurrence n. 重现,复现
interval n. 间歇,间隔

Estimating and Controlling Floods*

❶ Because people have **encroached** on the flood plains of many rivers, flooding is one of the most universally experienced geologic **hazards**. To minimize flood damage and loss of life, it is useful to know the potential size of large floods and how often they might occur. This is often a difficult task because of the lack of long-term records for most rivers. The U.S. Geological Survey[1] **monitors** the stage (water elevation) and **discharge** of rivers and streams throughout the United States to collect data that can be used to attempt to predict the size and frequency of flooding and to make estimates of water supply.

❷ **Hydrologists designate** floods based on their **recurrence interval**, or return period. For example, a 100-year flood is the largest flood expected to occur within a period of 100 years. This does not mean that a 100-year flood occurs once every century but

---

* The passage is adapted from PLUMMER C C, CARLSON D H, HAMMERSLEY L. Estimating the size and frequency of floods[M]//Physical geology. 15th ed. New York: McGraw-Hill Publishing, 2016: 256–259.

that there is a 1-in-100 chance, or a 1% probability, each year that a flood of this size will occur. Usually, flood control systems are built to **accommodate** a 100-year flood because that is the minimum **margin** of safety required by the **federal** government if an individual wants to obtain flood insurance **subsidized** by the Federal Emergency Management Agency (FEMA)[2].

❸ To calculate the recurrence interval of flooding for a river, the annual peak discharges (largest discharge of the year) are collected and ranked according to size (Table 1). The largest annual peak discharge is assigned a rank (m) of 1, the second a 2, and so on until all of the discharges are assigned a rank number. The recurrence interval (R) of each annual peak discharge is then calculated by adding 1 to the number of years of record (n) and dividing by its rank (m).

$$R = \frac{n+1}{m}$$

accommodate $v.$ 顺应,容纳,提供空间
margin $n.$ 边缘,余地,界限
federal $n.$ 联邦的,联邦政府的
subsidize $v.$ 资助,补助,给……发津贴

Table 1 Annual peak discharges and recurrence intervals in rank order for the Cosumnes River at Michigan Bar, California

Year	Peak Discharge/cfs	Magnitude Rank/m	Recurrence Interval/years
1997	93 000	1	100
1907	71 000	2	50
1986	45 100	3	33.33
1956	42 000	4	25.0
1963	39 400	5	20.0
1958	29 300	10	10.0
1928	22 900	20	5.0
1914	18 200	30	3.33
1918	11 900	40	2.5
1910	9 640	50	2.0
1934	7 170	60	1.67

❹ For example, the Cosumnes River in California has 99 years of record ($n=99$), and in 1907, the second-largest peak discharge

unreasonably *adv.* 不合理地
levee *n.* 防洪堤,码头
curve *n.* 曲线,弧线
plot *v.* 绘制(图表),(在地图上)画出
project *v.* 预估,设计,规划
slope *n.* 斜坡,坡度,坡地
dramatic *adj.* 巨大的

($m=2$) of 71,000 cfs$^{(3)}$ occurred. The recurrence interval (R), or expected frequency of occurrence, for a discharge this large is 50 years:

$$R = \frac{99+1}{2} = 50(\text{year})$$

That is, there is a 1-in-50, or 2%, chance each year of a peak discharge of 71,000 cfs or greater occurring on the Cosumnes River.

❺ The flood of record (largest recorded discharge) occurred on January 2, 1997, when heavy, **unseasonably** warm rains rapidly melted snow in the Sierra Nevada and caused flooding in much of northern California. A peak discharge of 93,000 cfs in the Cosumnes River resulted in **levee** breaks and widespread flooding of homes and agricultural areas. The recurrence interval for the 1997 flood (93,000 cfs) is 100 years:

$$R = \frac{99+1}{1} = 100(\text{years})$$

❻ A flood-frequency **curve** can be useful in providing an estimate of the discharge and the frequency of floods. The flood-frequency curve is generated by **plotting** the annual peak discharges against the calculated recurrence intervals (Figure 1). Because most of the data points defining the curve plot in the lower range of discharge and recurrence interval, there is some uncertainty in **projecting** larger flood events. Two flood-frequency curves are drawn in Figure 2; the red line represents the best-fit curve for all of the data, whereas the dashed blue line excludes the 1997 flood of record. Notice that the curve has a steeper **slope** when the 1997 data is included and that the size of the 100-year flood has increased from 73,000 cfs to 93,000 cfs based on the one additional year of record. Because large floods do not occur as often as small floods, the rare large flood can have a **dramatic** effect on the shape of the flood-frequency curve and the estimate of a 100-year event. This is particularly true for a river like the Cosumnes that has had only two large events, one in 1907 and the other 90 years later in 1997.

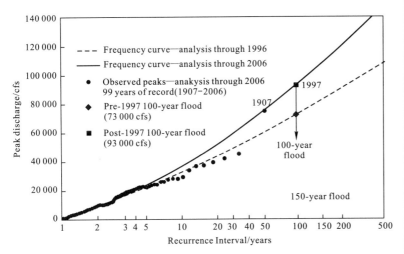

Figure 1 Flood-frequency curves for the Cosumnes River

Source: Data from Richard Hunrichs, hydrologist, U.S. Geological Survey and U.S. Geological Survey Water-Data Report, CA-97-3

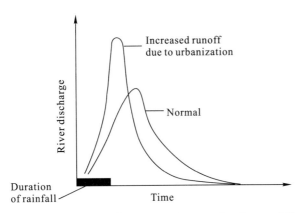

Figure 2 The presence of a city can increase the chance of floods

Analysis: One curve shows the normal increase in a river's discharge following a rainstorm (black bar). The other curve shows the great increase in runoff rate and amount caused by pavement and storm sewers in a city.

❼ The 100-year flood plain is based on the estimate of the discharge of the 100-year flood and on careful mapping of the flood plain. Changes in the estimated size of the 100-year flood could result in property that no longer has 100-year flood protection. In this case, property owners may be prevented from getting flood insurance or money to rebuild from FEMA unless new flood-control structures are built to provide additional protection or houses are raised or even **relocated** out of the flood plain.

paved area 铺筑过的区域
sewer n. 排水管,污水管
runoff n. 径流量,溢出
inhibiting adj. 起抑制作用的,抑制的,约束的
flashy adj.（因巨大等）显眼的
cloudburst n.（骤然降下的）大暴雨
overwhelm v. 淹没,压垮
outlet n. 出口,排放管
dock n. 码头,船坞
erosion n. 腐蚀
periodically adv. 周期性地,定期地
overtop v. 高出,超出
partially adv. 部分地
upstream adj. 上游的
dam n. 水坝
artificial adj. 人工的,人造的
embankment n. 堤岸,堤围
riprap n.（防冲）乱石,（防冲）乱石筑成的地基或堤坝
concrete n. 混凝土,水泥
floodwall n. 防洪墙,防洪堤
downstream n. 下游
bypass n. 旁道,旁通管
divert n. 使转向,使绕道,转移
weir n. 堰,拦河坝,导流坝
disastrous adj. 灾难性的
tributary n.（流入大河或湖泊的）支流

❽ Urbanization contributes to severe flooding. **Paved areas** and storm **sewers** increase the amount and rate of surface **runoff** of water. This is due to their **inhibiting** infiltration of rainwater into the ground and their rapid delivery of the resulting increased runoff to the channels, making river levels higher during storms. Such rapid increases in runoff or discharge to a river are called a "**flashy**" discharge. Storm sewers are usually designed for a 100-year storm; however, large storms that drop a lot of rain in a short period of time (**cloudburst**) may **overwhelm** sewer systems and cause localized flooding. Rising river levels may block storm sewer **outlets** and add to localized flooding problems. Bridges, **docks**, and buildings built on flood plains can also constrict the flow of floodwaters, increasing the water height and velocity and promoting **erosion**.

❾ The nature of all rivers is to **periodically** flood and **overtop** their levees. Wise land-use planning should go hand in hand with flood control. Wherever possible, buildings should be kept out of areas that might someday be flooded by 100-year floods.

❿ The risk of flooding to river cities may be **partially** reduced by engineered flood-control structures. **Upstream** dams can trap water and release it slowly after the storm. (A **dam** also catches sediment, which eventually fills its reservoir and ends its life as a flood-control structure.) **Artificial** levees are **embankments** built along the banks of a river channel to contain floodwaters within the channel. Protective walls of stone (**riprap**) or **concrete** are often constructed along river banks, particularly on the outside of curves, to slow erosion. **Floodwalls**, walls of concrete, may be used to protect cities from flooding; however, these flood-control structures may constrict the channel and cause water to flow faster with more erosive power **downstream**. **Bypasses** are also used along the Mississippi and other rivers to reduce the discharge in the main channel by **diverting** water through gates or **weirs** into designated basins in the flood plain. The bypasses serve to give part of the natural flood plain back to the river. Dams and levees are designed to control certain specified floods. If the flood-control structures on your river were designed for 75-year floods, then a much larger 100-year flood will likely overtop these structures and may destroy your home. The **disastrous** floods along the Missouri and Mississippi rivers and their **tributaries** north of Cairo, Illinois, in 1993 resulted from many such failures in flood control.

Unit 8 Natural Disasters

Notes

(1) **The U.S. Geological Survey** is the nation's largest water, earth, and biological science and civilian mapping agency. It collects, monitors, analyzes, and provides scientific understanding of natural resource conditions, issues, and problems.

(2) **The Federal Emergency Management Agency (FEMA)** in the United States supports citizens and emergency personnel to build, sustain, and improve the nation's capability to prepare for, protect against, respond to, recover from, and mitigate all hazards.

(3) **CFS/cfs** (cubic feet per second) is the standard unit of measurement of the flow of water. It is equivalent to about 2 m³ per minute.

Reading Comprehension

  Keys

Directions   Match the paragraphs with their appropriate subtitles.

1. Paragraph 1 _____   A. The merits and demerits of engineered flood-control structures

2. Paragraph 2 _____   B. The formula of recurrence interval of flooding

3. Paragraph 3 _____   C. The imprecision of large-floods prediction

4. Paragraph 4 _____   D. The link between changes in estimated size and property loss

5. Paragraph 5 _____   E. The accommodation criteria of flood controlling system

6. Paragraph 6 _____   F. Keep a distance from the potentially flood-affected area

7. Paragraph 7 _____   G. The estimated recurrence interval of the flood in 1997, California

8. Paragraph 8 _____   H. The necessity of knowing floods size and frequency

9. Paragraph 9 _____   I. How does urbanization increase the risk of flooding

10. Paragraph 10 _____   J. The exemplification of the recurrence interval formula

Passage B

Volcano Monitoring*

❶ As one of the most **destructive** occurrences on the Earth, volcanic eruptions are of great concern to people everywhere. The magma and ash that spew out of a volcano can cause much damage, posing an immediate threat to communities near it and causing extensive environmental and financial damages, too. For all of these reasons, scientists have developed several methods of monitoring volcanoes in order to catch any activity that foreshadows an eruption. The kinds or activity that are monitored are seismic disruptions, **deformations** in the volcano and landscape, and **gaseous** emissions from the magma itself. A wide range of techniques and equipment are used in the process.

❷ One of the most reliable and frequently employed methods of monitoring volcanoes and predicting eruptions is through the measurement of local seismic **disruptions**. This is because **impending** volcanic eruptions frequently **give off** minor earthquakes. As magma rises into the volcano, it must squeeze through a constricted **chamber** or series of chambers while under great pressure. The force will **crack** some of the rock or force rocks with pre-existing cracks to **convulse**, setting off **tremors** with varying frequencies. This seismic activity is relatively weak and occurs roughly ten kilometers beneath the volcano. In order to detect such weak quakes, scientists must set up a network of **seismometers** around the volcano in order to catch the slightest fluctuations in seismic activity that are usually a **prelude** to a volcanic eruption. About four to eight seismometers are placed about twenty kilometers from one of the vents, and several more are placed on the volcano itself. All of these have to be close in order to catch the quakes, as a seismometer being placed too far away could not detect subtle shifts in seismic activity. Fortunately, this method has been used so extensively that it is

---

* This passage is adapted from 语言研究中心, 2011. 托福官方指南考点详解：基础篇（阅读分册）[M]. 北京：北京语言大学出版社.

quite advanced, and scientists are experienced in detecting crucial seismic activity quickly, accurately, and in real time.

❸ While seismic monitoring is the most widely used and trusted method of monitoring volcanoes, there are other technologies that allow scientists to observe landscape deformation, which usually accompanies volcanic activity. In addition to causing quakes, the increased magma flow will also make the volcano swell and alter the surrounding landscape, particularly its evenness and elevation, though these changes are too slight to be noticed with the naked eye. Scientists use a variety of tools to observe such changes. One major tool is the satellite. In particular, scientists use the global positioning system (GPS) satellites to study a very detailed map of the Earth's surface, though the GPS is not primarily used to study volcanoes. Additionally, scientists have Interferometric Synthetic Aperture Radar (InSAR) satellites. These satellites use radar to map out changes in the landscape and the development of deformations very accurately, detecting possible volcanic activity. Changes on the ground can also be measured directly by people with land surveying equipment. Scientists use devices such as electronic distance meters (EDM) and **tiltmeters** to directly observe subtle changes that magma flow causes in the landscape. One way that scientists use all of these methods in combination is by **pinpointing** two distinct spots on a volcano with GPS satellites. They then use EDMs, tiltmeters, and InSAR satellites to measure the exact distance between them. As the volcano fills with magma, it stretches like a balloon and thus increases the distance between the two spots. If surveying tools like tiltmeters and EDMs catch this development, scientists can be **altered** to the possibility of an eruption before it occurs.

❹ In addition to the deformations in the ground caused by volcanic activity, scientist can measure chemical gaseous emission. Gauging the emission of certain gases like **sulfur dioxide** and carbon dioxide is helpful. As magma rises to the surface, it will give off greater amounts of these two gases, so increased amounts of them around a volcanic area would be a good indication of increased activity. While it is possible to monitor such activity at a safe distance via satellite, weather can interfere with accurate readings, so the direct sampling of these emissions by people is a more accurate method, though this means having to get near an active

tiltmeter *n.* 测斜器,地面倾斜度测量仪
pinpoint *v.* 确定,精准定位
alter *v.* 更改,修改
sulfur dioxide 二氧化硫

retrieve v. 找回,取回,挽回,索回

dissolve v. 溶解,消除

skew v. 偏离,歪斜

vanish v. 消失

vent to **retrieve** the samples. This is difficult because acidic gases like sulfur dioxide are easily **dissolved** in bodies of water, **skewing** precise measurements. Carbon dioxide is less likely to **vanish** in such a manner, though, so it could be more helpful in predicting volcanic activity.

Reading Comprehension

 Keys

Directions Answer the following following questions according to Passage B.

1. The word "foreshadows" in Paragraph 1 is closest in meaning to (   )
   A. follows after.
   B. may prevent.
   C. makes worse.
   D. comes before.

2. According to Paragraph 2, why must scientists use a network of seismometers around volcanoes? (   )
   A. The earthquakes that occur are not very strong.
   B. They must be ready to replace damaged instruments.
   C. The earthquakes are so strong that many are necessary.
   D. They are not certain where tremors will occur.

3. Which of the sentences below best expresses the essential information in the highlighted sentence? (   )
   A. Before an eruption, the land elevation and evenness around an earthquake become distorted.
   B. Magma flow transforms the land in ways that are hard to detect, as well as causing quakes.
   C. The increased flow of magma causes earthquakes around a volcano that are barely noticeable.
   D. The increased swelling of a volcano about to erupt is due to the earthquakes that shift land elevation.

4. According to Paragraph 3, tiltmeters help scientists predict volcanoes by (   )
   A. measuring the change in temperatures.
   B. measuring the emission of gases.
   C. measuring the evenness of the ground.
   D. measuring the fracturing of rocks.

5. The word "gauging" in Paragraph 4 is closest in meaning to (   )
   A. preventing.

B. causing.

C. measuring.

D. discovering.

6. The phrase "such activity" in Paragraph 4 refers to ( )

   A. the flowing of magma.

   B. the emission of gases.

   C. the monitoring of gases.

   D. deformations in the ground.

7. According to Paragraph 4, why are gaseous emissions from magma directly sampled? ( )

   A. Because accurate readings of emissions can be hindered by bad weather.

   B. Because it is impossible to read gas emissions using satellites.

   C. Because direct sampling of gases is the safest method.

   D. Because sampling of gases absorbed in the water bodies is required.

8. According to Paragraph 4, why is measuring sulfur dioxide emissions to monitor volcanic eruptions tricky? ( )

   A. Sulfur dioxide emissions are tainted carbon dioxide.

   B. Magma does not always release increased amounts of sulfur dioxide.

   C. Stormy weather can scatter sulfur dioxide gases away from volcanoes.

   D. Sulfur dioxide tends to dissolve in nearby bodies of water.

9. The letters [A], [B], [C], and [D] indicate where the following sentence (in bold) could be added to the following part of the passage. Where would the sentence best fit? ( )

   Both of these types of satellites provide scientists with the advantage of measuring such changes from a safe distance away from volcanic activity.

   These satellites use radar to map out changes in the landscape and the development of deformations very accurately, detecting possible volcanic activity. [A] Changes on the ground can also be measured directly by people with land surveying equipment. [B] Scientists use devices such as electronic distance meters (EDM) and tiltmeters to directly observe subtle changes that magma flow causes in the landscape. [C] One way that scientists use all of these methods in combination is by pinpointing two distinct spots on a volcano with GPS satellites. [D] They then use EDMs, tiltmeters, and InSAR satellites to measure the exact distance between them. As the volcano fills with magma, it stretches like a balloon and thus increases the distance between the two spots. If surveying tools like tiltmeters and EDMs catch this development, scientists can be altered to the possibility of an eruption before it occurs.

10. An introductory sentence (in bold) for a brief summary of the passage is provided below. Complete the summary by selecting THREE answer choices that express the most important ideas in the passage. ( )

> Scientists have developed several ways of monitoring different natural phenomena that usually indicate that a volcano is about to erupt.

A. The gases that magma gives off as it rises to the surface are measured closely or at a distance, though this method presents many challenges.

B. Scientists constantly look out for the occurrence of small earthquakes and tremors in the area around an active volcano.

C. Some gases from magma are tricky to measure because they disappear into surrounding lakes and rivers, making readings imprecise.

D. The use of GPS satellites is very effective in measuring seismic activity, but the financial expense is too great to use them often.

E. Many different instruments, including satellites and surveying techniques, are used to detect changes in the landscape before an eruption.

F. The earthquakes that occur due to increased volcanic activity are usually beneath the surface and are very weak.

Passage C

Scan and read along the passage

Mitigation of the Impacts on the Earth by Near-Earth Objects *

Scan and read along the vocabulary

mitigation *n.* 减缓,缓解
impact *n.* 撞击,冲击力
inevitable *adj.* 不可避免的
impactor *n.* 冲撞器
airburst *n.* 空爆,气压炸裂
inevitability *n.* 必然性,不可避免

❶ Impacts on the Earth by near-Earth objects (NEOs)[1] are **inevitable**. The **impactors** range from harmless fireballs, which are very frequent; through the largest **airbursts**, which do not cause significant destruction on the ground, on average occurring once in a human lifetime; to globally catastrophic events, which are very unlikely to occur in any given human lifetime but are probably randomly distributed in time. The risks from these NEOs, or more specifically scientists' assessment of the risks in the next century, will be changing as surveys are carried out. Given the **inevitability** of impacts, and noting that the entire point of surveys is to enable appropriate action to be taken, how can the effects of potential impacting NEOs be mitigated?

❷ In this part the committee considers four categories of mitigation:

---

* This passage is adapted from National Academy of Sciences, 2010. Defending planet earth: near-earth-object surveys and hazard mitigation strategies [R]. Washington, D.C.: The National Academies Press.

- **Civil defense**—involving such efforts as **evacuating** the region around a small impact.
- Slow-push-pull techniques—gradually changing the orbit of a NEO so that it misses the Earth.
- Kinetic impact—delivering a large amount of **momentum** (and energy) **instantaneously** to a NEO to change its orbit so that it misses the Earth.
- Nuclear **detonation**—delivering a much larger amount of momentum (and energy) instantaneously to a NEO to change its orbit so that it misses the Earth.

Civil defense: disaster preparation and recovery

❸ Of the two generic approaches to the mitigation of an impact hazard—1) active **orbital** change or the destruction of the incoming body, and 2) passive, traditional natural-disaster mitigation based on "all-hazards" **protocols** for evacuation, sheltering, response and recovery, and so on—people in contemporary society would very likely be faced with evacuation and sheltering rather than orbital change or destruction during their lifetimes. The most probable event will be a very late warning of a small NEO (tens of meters in diameter or less). At the opposite end of the size **spectrum** for impacts approaching or exceeding the level of "civilization-threatening impacts" (100 to many hundreds of meters in diameter), there are inadequate **precedents**.

Slow-push-pull techniques

❹ This section considers the first of three approaches to prevent an impact rather than to protect against an impact. "Slow-push-pull" means the continuous application of a small but steady force to the NEO, thereby causing a small acceleration of the body relative to its **nominal** orbit. The effect of such small accelerations is most productive if applied along or against the NEO's direction of motion, as this causes a net shift of the NEO along its orbit. This shift can **avert** an impact by causing the NEO to "show up" at the Earth's orbit earlier or later than the Earth does. A simple **rule-of-thumb** formula predicts the drift along the NEO's orbit for a given applied acceleration

$$\Delta s = \pm \frac{3}{2} A t_a (t_a + 2 t_c)$$

civil defense 民防
evacuate *v.* 撤离,排空
momentum *n.* 动量,动力
instantaneously *adv.* 瞬间地,立即
detonation *n.* 爆炸
orbital *adj.* (行星或空间物体)轨道的
protocol *n.* 议定书,条约草案
spectrum *n.* 范围
precedent *n.* 先例,可援用参考的具体例子
nominal *adj.* 名义上的,象征性的
avert *v.* 防止,避免,阻止
rule-of-thumb *n.* 靠经验估计,依据经验的方法

where Δs represents the shift in the NEO's position relative to its nominal orbit, A represents the induced acceleration of the NEO, t_a represents the time during which the acceleration is applied, and t_c represents the coast time after the application of the acceleration. For estimating the range of NEOs for which a given method is applicable, the committee considers orbital changes large enough to move the NEO by 15,000 kilometers, enough to provide a safe miss as long as the orbit is well determined. Assume that a 10-ton spacecraft is the maximum possible with current launch capability and that a 50-ton spacecraft might be possible with future heavy-lift launch vehicles. Of course, multiple launches are possible and may be desirable to **scale up** the effect or to provide **backup** in case of failure.

❺ Slow-push-pull techniques are the most accurately controllable and are adequate for changing the orbits of small NEOs (tens of meters to roughly 100 meters in diameter) with decades of advance warning and for somewhat larger NEOs (hundreds of meters in diameter) in those few cases in which the NEO would pass through a **keyhole** that would put it onto an impact **trajectory**. Of the slow-push-pull techniques, the **gravity tractor** appears to be the most independent of variations in the properties of the NEO and by far the closest to technological readiness.

Kinetic impact methods

❻ Kinetic impact mitigation uses one or more very-high-velocity (typically more than about 5 km/s) impacts of a large spacecraft ("impactor") into a hazardous object. These impacts would change the velocity of the hazardous object by some small amount, which would result in a new orbit for the hazardous object that would cause it to miss the Earth. The method is relatively simple and effective for NEOs with diameters up to about half a kilometer, and it is well within current capabilities given modest hardware and control developments. This method would likely be the method of choice for the mitigation of hazardous objects of the size range just indicated when there are years or more of warning time.

❼ In this approach either the spacecraft can "run into" the hazardous object, or the hazardous object can run into the spacecraft; only the relative velocity of the impact is relevant. The achievable relative velocity varies significantly with the details of the NEO's orbit, but, unlike the variability in other parameters that affect this and other methods, the orbit of any particular NEO will be known with sufficient **accuracy** that various spacecraft trajectories can be studied with a view to achieving the maximum relative velocity in the best direction at encounter. NASA's Deep Impact mission[2] in 2006 demonstrated this principle, although with a smaller impactor on a larger body (6 kilometers in diameter). That impact was at 10 km/s and the committee will adopt that value for estimating effectiveness, but it is noted that for present capabilities the range of relative velocities due to different orbits of the NEOs is likely to be anywhere from a few to a few tens of kilometers per second.

Nuclear methods

❽ Nuclear explosives constitute a mature technology, with well-characterized **outputs**. They represent by far the most **mass-efficient** method of energy transport and should be considered as an option for NEO mitigation. Nuclear explosives provide the only option for large NEOs (>500 meters in diameter) when the time to impact is short (years to months), or when other methods have failed, and time is running out. The extensive test history of nuclear explosives demonstrates a proven ability to provide a **tailored** output (the desired mixture of X rays, neutrons, or **gamma rays**) and **dependable yields** from about 100 tons to many **megatons** of TNT-equivalent energy. **Coupled with** this test history is an abundance of data on the effects of surface and subsurface **blasts**, including shock generation and **cratering**.

❾ Various methods have been proposed for using nuclear explosions to reduce or eliminate a NEO threat; for a given mass of the NEO, the warning time is a primary criterion for choosing among them. With decades of warning, the required change in velocity (ΔV) from the explosion is millimeters to a centimeter per second and can be met for NEOs several kilometers in diameter. This range of values is much less than the 25 to 50 cm/s escape velocity from **moderate** to large (500- to 1,000-meter-diameter) bodies, so it is reasonable to assume that such a small ΔV would not lead to the target's **fragmentation** or to excessive **ejecta** (i.e., debris thrown off the object). In models of NEOs with surface

accuracy *n.* 准确度,精度
output *n.* 产出
mass-efficient *adj.* 质量效率高的,即轻质量可以获得大能量的
tailored *adj.* 专门设定的
gamma rays 伽马射线
dependable *adj.* 可信赖的,可靠的
yield *n.* 产出,收益
megaton *n.* 百万吨级(爆炸能量计量单位,相当于100万吨黄色炸药的威力),兆吨
coupled with 加上,加之
blast *n.* 爆炸
cratering *n.* (被炸成的)坑,孔
moderate *adj.* 中等的,适度的
fragmentation *n.* 破碎,破裂
ejecta *n.* 喷出物,排出物

porosity *n.* 孔隙度,孔隙率
dissipative *adj.* 消散的,消耗(性)的
assembly *n.* 组装
fuse *n.* 导火线
container *n.* 容器
cylinder *n.* 气缸
incorporate *v.* 包含,吸收,将……包括在内,使并入
constraint *n.* 约束(条件),限制
specification *n.* 规范,规格
latency *n.* 延迟,潜伏(期)

densities as in terrestrial environments, nearly 98 percent of a body remains bound as a single object through only its own weak gravity. The small amount of ejecta expands over the decades to form a large cloud of low-density debris, reducing its posed threat by another factor of 104 to 105. The amount of the ejecta depends on the surface **porosity**. As in the case of kinetic impacts, a **dissipative**, low-density surface will reduce the amount of ejecta, thus reducing the ΔV.

❿ Alternatively, when the time to projected impact is short (i.e., years rather than decades), it may be impossible to apply a sufficient ΔV without fragmentation, but the limiting factor is **assembly** and launch. A nuclear package with a new **fuse** (i.e., a fuse that is not designed for terrestrial use) and a new **container** requires a **cylinder** about 1 meter in length and 35 centimeters in diameter, with a mass under 220 kilograms. The longest lead-time item for **incorporating** such a device in a rocket system is the development of a container to deliver the device and a fusing system capable of operating with the timing **constraints** required by the spacecraft velocities near impact with the NEO. **Specifications** for a nuclear bus could be the same as those for a kinetic-impactor mission, but it would be very challenging to construct and integrate with the booster rocket and the nuclear package in under a year. This "**latency** time" between the decision to act and the launch can be reduced dramatically (perhaps as much as 100-fold) by designing and testing these critical components in advance of discovering a hazardous NEO.

Notes

(1) **Near-Earth Objects (NEOs)** are comets and asteroids that have been nudged by the gravitational attraction of nearby planets into orbits that allow them to enter the Earth's neighborhood.

(2) **NASA**'s Deep Impact mission was the first attempt to peer beneath the surface of a comet. On July 4, 2005, the Deep Impact spacecraft delivered a special impactor into the path of comet Tempel 1 to reveal never before seen materials and provide clues about the internal composition and structure of a comet.

Unit 8　Natural Disasters

Reading Comprehension

Directions　Fill in the blanks according to Passage C.

NEOs Impacts Mitigation Methods	Best Application Condition
civil defense	a very 1._____ warning of a 2._____ NEO (≤3._____ in diameter)
slow-push-pull techniques	orbital changes large enough to move the NEO by 4._____ kilometers
kinetic impact method	for NEOs with diameters up to about 5._____
nuclear method	6._____ NEOs (7._____ meters in diameter) when the time to impact is 8._____ (years to months), or when other methods have 9._____ and time is 10._____

Exercises

Vocabulary

Ⅰ. Technical words

Directions　Write the English expressions according to the given Chinese.

防洪墙　　　　_____

表面径流　　　_____

测斜仪　　　　_____

近地天体　　　_____

撞击;冲击力　　_____

核爆炸　　　　_____

轨迹　　　　　_____

中子,中子射线　_____

Ⅱ. Academic vocabulary

Directions　Fill in the blanks with appropriate forms of the verbs given below. Each word can be used only once.

Verb	Definition in Academic Writing
avert	to prevent the occurrence of; to turn away or aside
convulse	to shake uncontrollably; to move or stir about violently
designate	to assign a name or title to
encroach	to advance beyond the usual limit; to impinge or infringe upon
evacuate	to move out of an unsafe location into safety; to empty completely
incorporate	to make into a whole or make part of a whole; to include or contain
induce	to cause something; to call forth or bring about by influence or stimulation
mitigate	to reduce the harmful effects of something
pinpoint	to locate or aim with great precision or accuracy
subsidize	to support grants of (often government) money

1. Efforts to _____ its location have been difficult in the vast Atlantic and with no communication from the ship's 15-member crew.
2. In this case it is desirable to _____ additional oxidizer into the composition.
3. We _____ the study of the Earth using physical measurements at the surface as geophysics.
4. Improved surface water drainage was the only method used to _____ rainfall damages.
5. The inhabitants were _____ from the flooded village.
6. The island was _____ by an earthquake.
7. The advertisement is meant to _____ people to eat more fruit.
8. The government has _____ the car industry.
9. The dyke will be constructed to _____ floods.
10. The sea is gradually _____ on the land.

Sentence Structure

Understand passive voice in English sentences.

In academic context, passive voice sentences are common. The reason is that the focus of academic writing is on intellectual ideas and factual information instead of on personal emotions or individual experiences. Thereby, the authors tend to avoid personal expressions out of intuition, feeling, prejudice or one's own experience.

In these sentences, the passive voice is appropriate. That is to say, passive sentence structure is suitable in the following situations:

- The subject/doer/agent is unknown, irrelevant, or very general and obvious (e.g., an experimental solar power plant will be built…; several attempts *have been made* to…), also

as is shown in the following 1st and the 2nd sentences;
- Talking about a general truth (e.g., *It is generally assumed/accepted* ...), also as is shown in the following 3rd sentence;
- Emphasizing the person or thing acted on (e.g., *a catheter was inserted*), also as is shown in the following 4th sentence.

For example,

An experimental solar power plant *will be built* in the Australian desert.

Several attempts *have been made* to explain the phenomenon.

It is generally assumed that the Dravidians were the original inhabitants of South Asia and the Aryans displaced them.

A catheter was inserted for post-operative bladder irrigation.

However, overuse of passive voice can cause confusion, wordiness, and indirectness, which will ultimately make the reader work unnecessarily hard. As a result, the past several years has witnessed a declining usage of passive voice within many science disciplines. Please avoid overusing passive voice sentences in your writing.

Directions Analyze the following sentences by underlining the passive voices used in these sentences.

1. Usually, flood control systems are built to accommodate a 100-year flood because that is the minimum margin of safety required by the federal government if an individual wants to obtain flood insurance subsidized by the Federal Emergency Management Agency (FEMA). (Paragraph 2, Passage A)

2. Fortunately, this method has been used so extensively that it is quite advanced, and scientists are experienced in detecting crucial seismic activity quickly, accurately, and in real time. (Paragraph 2, Passage B)

3. That impact was at 10 km/s, and the committee will adopt that value for estimating effectiveness, but it is noted that for present capabilities the range of relative velocities due to different orbits of the NEOs is likely to be anywhere from a few to a few tens of kilometers per second. (Paragraph 7, Passage C)

4. Various methods have been proposed for using nuclear explosions to reduce or eliminate a NEO threat; for a given mass of the NEO, the warning time is a primary criterion for choosing among them. (Paragraph 9, Passage C)

Translation

I. Translate the following English sentences into Chinese. Pay attention to the passive voice in each sentence.

1. Usually, flood control systems are built to accommodate a 100-year flood because that is the minimum margin of safety required by the federal government if an individual wants to obtain flood insurance subsidized by the Federal Emergency Management Agency (FEMA).

2. Fortunately, this method has been used so extensively that it is quite advanced, and scientists are experienced in detecting crucial seismic activity quickly, accurately, and in real time.

3. That impact was at 10 km/s, and the committee will adopt that value for estimating effectiveness, but it is noted that for present capabilities the range of relative velocities due to different orbits of the NEOs is likely to be anywhere from a few to a few tens of kilometers per second.

4. Various methods have been proposed for using nuclear explosions to reduce or eliminate a NEO threat; for a given mass of the NEO, the warning time is a primary criterion for choosing among them.

II. Translate the following Chinese paragraph into English.

近几年,频发的自然灾害让全球深受其害。炎热的酷暑、狂暴的飓风、刺骨的严寒以及滔天的洪水近乎成了常态。面对日益脆弱的全球气候,人类需要更认真地思考如何切实有效地规范自身活动,珍爱我们共同的家园。同时,未来极端灾害天气可能对水利、农业、林业、能源、健康、旅游等相关领域的影响更大,世界经济复苏将面临更多的不确定性。

Glossary

A

abiotic /ˌeɪbaɪˈɑːtɪk/	4C
adj. 非生物的,无生命的	
abrasion /əˈbreɪʒn/	2B
n. 磨损,擦伤	
accommodate /əˈkɑːməˌdeɪt/	8A
v. 顺应,容纳,提供空间	
accumulate /əˈkjuːmjəleɪt/	3A
v. 积累,积攒	
accumulation /əˌkjuːmjəˈleɪʃn/	2A
n. 积聚	
accuracy /ˈækjərəsi/	8C
n. 准确度,精度	
acidic /əˈsɪdɪk/	3A
adj. 酸性的	
adiabatically /ˌædiəˈbætɪkəli/	1C
adv. 绝热地	
adversely /ˈædvɜːrsli/	3C
adv. 不利地,有害地	
aeroponics /ˌeərəʊˈpɒnɪks/	4B
n. 空气种植法	
aerosol /ˈeərəsɑːl/	3A
n. 气溶胶	
aesthetic /esˈθetɪk/	2C
adj. 审美的,美学的	
afforestation /əˌfɔːrɪˈsteɪʃn/	7A
n. 造林	
aggregates /ˈægrɪgeɪts/	2C
n. 混凝土	
airburst /ˈeərbɜːrst/	8C
n. 空爆,气压炸裂	
albedo /ælˈbiːdoʊ/	3B
n. 反照率,反射率	
albite /ˈælbaɪt/	2A
n. 钠长石	
algae /ˈældʒiː/	5C
n. 水藻,海藻	
align /əˈlaɪn/	4C
n. 公开支持,与……结盟,使平行,加入	
alkaline /ˈælkəlaɪn/	1C
adj. 碱性的,含碱的	
alpine /ˈælpaɪn/	4B
adj. 阿尔卑斯山的,高山的	
alter /ˈɔːltər/	8B
v. 更改,修改	
altitude /ˈæltɪtuːd/	3A
n. 海拔高度	
aluminum /əˈluːmɪnəm/	4B
n. 铝	
amber /ˈæmbə(r)/	2A
n. 琥珀	
ambient /ˈæmbiənt/	6C
adj. 环境的,周围的	
amend /əˈmend/	2A
v. 修改,修订	
amethyst /ˈæməθɪst/	2B
n. 紫水晶	

amplitude /ˈæmplɪtuːd/	3C	arsenic /ˈɑːrsnɪk/	4C
n. 广度, 阔度		*n*. 砷, 砒霜	
anemone /əˈnemənɪ/	5C	artificial /ˌɑːrtɪˈfɪʃl/	8A
n. 银莲花, 银莲花属, 海葵(sea anemone)		*adj*. 人工的, 人造的	
anomalous /əˈnɑːmələs/	3C	aseismic ridge /eɪˈzaɪzmɪk rɪdʒ/	1C
adj. 异常的, 不规则的		无震海岭	
anomaly /əˈnɑːməlɪ/	3C	ashlar /ˈæʃlər/	2C
n. 异常事物, 反常现象		*n*. 琢石	
anorthite /ænˈɔːθaɪt/	2A	asphalt /ˈæsfɔːlt/	3B
n. 钙长石		*n*. 沥青, 柏油	
Antarctica /ænˈtɑːrktɪkə/	5B	assembly /əˈsemblɪ/	8C
n. 南极洲		*n*. 组装	
anthocyanin /ˌænθəˈsaɪənɪn/	4B	asthenosphere /æsˈθɪnəˌsfɪr/	1C
n. 花色素苷		*n*. 软流圈, 岩流圈	
Anthropocene /ˈænθrəpəˌsiː/	1A	atmosphere /ˈætməsfɪr/	5A
n. 人类世		*n*. 大气, 大气层	
anthropogenic /ˌænθrəpəˈdʒenɪk/	3C	atmospheric gas /ˌætməsˈferɪk ˈɡæsz/	3A
adj. 人为的		*n*. 大气气体	
apatite /ˈæpətaɪt/	2A	atom /ˈætəm/	5C
n. 磷灰石		*n*. 原子, 微量, 极小量	
apparatus /ˌæpəˈrætəs/	2B	attendant /əˈtendənt/	6C
n. 设备, 器具		*adj*. 伴随的, 随之产生的	
appliance /əˈplaɪəns/	7B	attribute /ˈætrɪbjuːt/	2C
n. 家用电器, 装置		*n*. 属性, 特质	
aqueous /ˈeɪkwɪəs/	2A	augite /ˈɔːdʒaɪt/	2A
adj. 水的, 水般的		*n*. 辉石	
aquifer /ˈækwəfərz/	5A	avail /əˈveɪl/	6A
n. 地下含水层		*v*. 有帮助, 有益, 有用	
archaeological /ˌɑːrkɪəˈlɑːdʒɪkl/	2C	availability /əˌveɪləˈbɪlətɪ/	3C
adj. 考古学的, 考古的		*n*. 可用性, 可得性	
archaeologist /ˌɑːrkɪˈɑːlədʒɪst/	3C	avert /əˈvɜːrt/	8C
n. 考古学家		*v*. 防止, 避免; 阻止	
argon /ˈɑːrɡɑːn/	3A	axis /ˈæksɪs/	6B
n. 氩		*n*. 轴, 轴线, 对称中心线	

B

backup /ˈbækʌp/	8C
n. 后援，备份	
bacteria /bækˈtiriə/	5C
n. 细菌（bacterium 的复数）	
basin /ˈbeisn/	5C
n. 流域，盆地，海盆	
bedrock /ˈbedrɑːk/	4A
n. 基岩，基本原理	
benthos /ˈbenθɑs/	5C
n. 海底，海底生物，海底的动植物群	
binder /ˈbaindər/	2C
n. 黏合剂	
biodiversity /ˌbaioʊdaiˈvɜːrsəti/	1A
n. 生物多样性	
biogeochemical /baiədʒiəˈkemikl/	1A
adj. 生物地球化学的	
biophysical /baiəʊˈfizikəl/	3C
adj. 生物物理学的	
biosphere /ˈbaiəʊsfiə/	1A
n. 生物圈	
biota /baiˈoʊtə/	1A
n. 生物区（系），一时代（一地区）的动植物	
biotical /baiˈɑːtikəl/	4C
adj. 关于生命的，生物的	
biotite /ˈbaiəˌtait/	2A
n. 黑云母	
blast /blæst/	8C
n. 爆炸	
blocky stone /ˈblɑki stoʊn/	2C
块状石头	
bombard /bɑːmˈbɑːrd/	6C
n. 轰炸，连环炮击，（物）以高速粒子轰击	
borehole /ˈbɔːrhoʊl/	7C
n. 钻孔，井眼	
bottom-dwelling /ˈbɑːtəm ˈdwelin/	5C
adj. 底栖生物的	
breakthrough /ˈbreikθruː/	7A
n. 突破，重大进展	
breeze /briːz/	3B
n. 微风	
bromide /ˈbroʊmaid/	5C
n. 溴化物	
budget /ˈbʌdʒit/	6B
n. 预算，（机构、政府等的）财政收支状况	
buffer /ˈbʌfər/	4C
n. 缓存，缓冲器	
v. 缓解，存储	
buildup /ˈbilˌdʌp/	3A
n. 积聚，组合	
bulge /bʌldʒ/	6B
n. 鼓起，凸出，骤增，暴涨	
bunker /ˈbʌŋkər/	6C
n. 沙坑，煤仓，燃料库	
burrow /ˈbɜːroʊ/	4A
v. 掘地洞，挖地道	
bypass /ˈbaipɑːs/	8A
n. 旁道，旁通管	
by-product /ˈbai prɑːdʌkt/	6C
n. 副产品，附带产生的结果	

C

cadmium /ˈkædmiəm/ 4B
 n. (化学元素)镉
calcite /ˈkælsait/ 2A
 n. 方解石
calcium /ˈkælsiəm/ 2A
 n. (化学元素)钙
canister /ˈkænistər/ 6C
 n. 筒,小罐,防毒面具的滤毒罐
carbon /ˈkɑːrbən/ 7A
 n. 碳
carbon dioxide /ˌkɑːbən daiˈɒksaid/ 5C
 二氧化碳
catastrophe /kəˈtæstrəfi/ 1A
 n. 灾难,灾祸,困境
cavern /ˈkævərn/ 7C
 n. 洞穴,凹处
ceramic /səˈræmik/ 6C
 adj. 陶瓷的
chamber /ˈtʃeimbər/ 8B
 n. 室;密闭空间
chloride /ˈklɔːraid/ 4B
 n. 氯化物
chlorination /ˌklɔːriˈneiʃn/ 5A
 n. 氯化作用,加氯消毒
chlorofluorocarbon /ˌklɔːrouˈfluroukɑːrbən/ 3A
 n. 氯氟化碳
chlorophyll /ˈklɔːrəfil/ 4B
 n. 叶绿素
chlorosis /kləˈrousis/ 4B
 n. 变色病,萎黄病
chronology /krəˈnɑːlədʒi/ 3C
 n. 年表,年代学
civil defense /ˈsivl diˈfens/ 8C
 民防(系统)
cladding /ˈklædiŋ/ 7A
 n. 包层,外墙

cleavage /ˈkliːvidʒ/ 2B
 n. 解理
climate variability /ˈklaimət ˌveriəˈbiləti/ 3C
 气候变异
climatological /ˌklaimətəˈlɑːdʒikl/ 1A
 adj. 与气候学有关的
cloudburst /ˈklaudbɜːrst/ 8A
 n. (骤然降下的)大暴雨
cobalt /ˈkoubɔːlt/ 4B
 n. 钴,钴类颜料
co-evolutionary /kouˌevəˈluːʃəneri/ 1A
 adj. 协同进化的
coincide with /ˌkəuinˈsaid wið/ 1C
 符合,与……相一致
combustion /kəmˈbʌstʃən/ 7A
 n. 燃烧,氧化
commodity /kəˈmɑːdəti/ 2B
 n. 商品,货物
compacted /ˈkɑːmpæktid/ 5B
 adj. 压实的,压紧的
compensating /ˈkɑːmpenseitiŋ/ 3B
 adj. 补偿的,平衡的
complementary /ˌkɑːmpliˈmentri/ 7C
 adj. 互补的,相辅相成的
composition /ˌkɑːmpəˈziʃn/ 2C
 n. 成分构成,成分
compost /ˈkɑːmpoust/ 4B
 n. 堆肥,混合物
 v. 堆肥,施堆肥
concentrate /ˈkɑːnsntreit/ 7A
 v. 集中注意力,聚集,浓缩
concentration /ˌkɒns(ə)nˈtreiʃ(ə)n/ 2A
 n. 集中,聚集,浓缩,含量,浓度
concrete /ˈkɑːŋkriːt/ 8A
 n. 混凝土,水泥
condensation /ˌkɑːndenˈseiʃn/ 5A
 n. 冷凝,凝结

condense /kənˈdens/	3B	converge /kənˈvɜːrdʒ/	3B
v. 冷凝, 凝结		v. (使)汇聚, 集中	
conductivity /ˌkɑːndʌkˈtivəti/	6C	convergence /kənˈvɜːrdʒəns/	1A
n. 传导性		n. 趋同, 汇集, 相交	
configuration /kənˌfɪɡjəˈreɪʃn/	2B	convulse /kənˈvʌls/	8B
n. 布局, 构造, 配置		v. 使震动(或抖动)	
consensus /kənˈsensəs/	7A	cost-effective /ˌkɔːst ɪˈfektɪv/	4C
n. 一致看法, 共识		adj. 有成本效益的, 划算的	
Conservation Biology	6A	coupled /ˈkʌpld/	3C
/ˌkɑːnsərˈveɪʃn baɪˈɑːlədʒi/		adj. 耦合的	
n. 保护生物学		coupled with /ˈkʌpld wɪð/	8C
conservationist /ˌkɑːnsərˈveɪʃənɪst/	6A	加上, 加之	
n. (自然环境、野生动植物等)保护主义者		crack /kræk/	8B
consistent /kənˈsɪstənt/	3C	v. 断裂; 破裂	
adj. 一致的		cratering /ˈkreɪtərɪŋ/	8C
constituent /kənˈstɪtʃuənt/	3A	n. (被炸成的)坑, 孔	
n. 成分, 构成要素		crop /krɑːp/	6A
constrain /kənˈstreɪn/	3C	v. 出露, 露头	
v. 限制, 约束		cryosphere /ˈkraɪəsfɪr/	3A
constraint /kənˈstreɪnt/	8C	n. 冰冻圈	
n. 约束(条件), 限制		culmination /ˌkʌlmɪˈneɪʃn/	5A
container /kənˈteɪnər/	8C	n. 终点, 高潮	
n. 容器		curve /kɜːrv/	2B; 8A
contaminant /kənˈtæmɪnənt/	5A	v. (使)沿曲线移动, 呈曲线形	
n. 污染物, 致污物		n. 曲线, 弯曲, 曲面	
continental shelf /ˌkɒntɪˌnentəl ˈʃelf/	5C	cyclone /ˈsaɪkloʊn/	3C
大陆架		n. 气旋, 龙卷风	
convection /kənˈvekʃn/	1C	cylinder /ˈsɪlɪndər/	8C
n. (热通过气体或液体的)运流, 对流		n. 气缸	

D

dam /dæm/	8A	decommissioning /ˌdiːkəˈmɪʃnɪŋ/	6C
n. 水坝		n. 退役, 停运, 停止运作	
debris /dəˈbriː/	3A	decomposed /diːˈkəmpəʊzd/	4A
n. 碎片, 残骸		adj. 分解的	
decipher /dɪˈsaɪfər/	3A	deduce /dɪˈduːs/	1C
v. 破译		v. 推论, 推断	

deforestation /ˌdiːˌfɔːriˈsteiʃn/ 6A
  n. 毁林, 滥伐森林
deformation /ˌdiːfɔːrˈmeiʃn/ 8B
  n. 变形
degradation /ˌdegrəˈdeiʃn/ 6A
  n. 退化, 恶化, 质量下降
demography /diˈmɑːgrəfi/ 6A
  n. 人口组成, 人口统计
dependable /diˈpendəbl/ 8C
  adj. 可信赖的, 可靠的
depleted /diˈpliːtid/ 7C
  adj. 不足的, 耗尽的
depletion /diˈpliːʃn/ 1A
  n. 损耗, 耗尽
depositional /ˌdepəˈziʃənəl/ 5A
  adj. 沉积作用的
desalination /ˌdiːˌsæliˈneiʃn/ 5A
  n. (海水的)脱盐, 淡化
desertification /diˌzɜːrtifiˈkeiʃn/ 6A
  n. (土壤)荒漠化, 沙漠化(等同于 desertization)
designate /ˈdezignət/ 8A
  v. 命名, 标示, 指定, 指派
destruction /diˈstrʌkʃn/ 2C
  v. 破坏, 摧毁
destructive /diˈstrʌktiv/ 8B
  adj. 破坏性的, 有害的
detonation /ˌdetəˈneiʃn/ 8C
  n. 爆炸
deuterium /duːˈtiriəm/ 6C
  n. 氘, 重氢
diagnostic /ˌdaiəgˈnɑːstik/ 2B
  adj. 判断的
diatom /ˈdaiətəm/ 5C
  n. 硅藻
diffuse /diˈfjuːs/ 7A
  adj. 扩散的, 弥漫的, 难解的, 冗长的
dike /daik/ 1C
  n. 堤坝, 沟

dimension /daiˈmenʃn/ 2C
  n. 大小, 尺寸
dimension stone /daiˈmenʃn stoʊn/ 2C
  规格石材
dioxide /daiˈɑːksaid/ 7A
  n. 二氧化物
disastrous /diˈzæstrəs/ 8A
  adj. 灾难性的
discharge /disˈtʃɑrdʒ/ 8A
  n. 排放
  v. 排出, 流出
disciplinary /ˈdisəplənəri/ 1A
  adj. 学科的
discrepancy /diˈskrepənsi/ 2A
  n. 差异, 不符
disintegrated /diˈsintigreitid/ 4A
  adj. 分解的
dispersion /diˈspɜːrʒn/ 3B
  n. 分散, 散布, 分布
displacement /disˈpleismənt/ 1C
  n. 移位
dispose /diˈspoʊz/ 7C
  v. 处理, 放置, 安排
disrupt /disˈrʌpt/ 3A
  v. 中断, 扰乱
disruption /disˈrʌpʃn/ 8B
  n. 分裂, 中断
dissimilar /diˈsimilər/ 4A
  adj. 不同的
dissipate /ˈdisipeit/ 6C
  v. (使某事物)消散, 消失, 挥霍, 耗费
dissipative /ˈdisipeitiv/ 8C
  adj. 消散的, 消耗(性)的
dissolve /diˈzɑːlv/ 8B
  v. 溶解, 消除
dissolved /diˈzɑːlvd/ 5C
  adj. 溶解的, 溶化的
distinctive /diˈstiŋktiv/ 2B
  adj. 独特的, 与众不同的

divert /daɪˈvɜːrt/	8A		driver /ˈdraɪvər/	6A
v. 使转向,使绕道,转移			n. 驱动程序,驱动因素	
dock /dɑːk/	8A		droplets /ˈdrɑːpləts/	3B
n. 码头,船坞			n. 小水滴	
domestic /dəˈmestɪk/	3B		droughts /draʊts/	7A
adj. 家用的,家庭的			n. 干旱	
downstream /ˌdaʊnˈstriːm/	8A		dub /dʌb/	2A
n. 下游			v. 把……称为	
downwind /ˌdaʊnˈwɪnd/	3B		durable /ˈdʊrəbl/	2C
adv. 顺风,在下风			adj. 持久的,耐用的	
dramatic /drəˈmætɪk/	8A		dwarf /dwɔːrf/	6B
adj. 巨大的,突然的,激动人心的			v. 使显得矮小,使相形见绌	
drastically /ˈdræstɪkli/	1B		dwelling /ˈdwelɪŋ/	2C
adv. 彻底地,激烈地			n. 住宅,住所	
drill /drɪl/	6A		dynamics /daɪˈnæmɪks/	1A
v. 钻(孔),打(眼)			n. 动力学,力学,动力	

E

ecology /iˈkɑːlədʒi/	1A		emissions /iˈmɪʃnz/	3B
n. 生态学			n. 排放,排放物,辐射	
ecosystem /ˈiːkoʊsɪstəm/	6A		emit /iˈmɪt/	3A
n. 生态系统			v. 排放,散发	
eject /iˈdʒekt/	6C		encompass /ɪnˈkʌmpəs/	4C
v. 喷射,排出,弹出			n. 包含,包括	
ejecta /iˈdʒektə/	8C		encroach (on) /ɪnˈkroʊtʃ/	8A
n. 喷出物,排出物			v. 侵蚀,蚕食(土地)	
El Niño /el ˈniːnjoʊ/	3C		engulf /ɪnˈɡʌlf/	5B
厄尔尼诺现象(热带东太平洋上每隔几年出现的一股暖流)			v. 吞没,淹没	
			enhance /ɪnˈhæns/	3C
electromagnetic /ɪˌlektroʊmæɡˈnetɪk/	3A		v. 增强,提高	
adj. 电磁的			ensemble /ɑːnˈsɑːmbəl/	1A
elevation /ˌelɪˈveɪʃn/	5B		n. 共同,一起,集合,整体	
n. 高度,海拔			enthalpy /enˈθælpi/	7C
embankment /ɪmˈbæŋkmənt/	8A		n. 焓,热函	
n. 堤岸,堤围			enthusiasm /ɪnˈθuːziæzəm/	7A
embody /ɪmˈbɑːdi/	2C		n. 热情,热忱	
v. 使具体化			entity /ˈentəti/	1A
			n. 实体,存在	

environmentalist /ɪnˌvaɪrənˈmentəlɪst/ 7A
　　n. 环境保护主义者,环境保护论者,环境艺术家
envisage /ɪnˈvɪzɪdʒ/ 6B
　　v. 想象,设想,展望
epitomize /ɪˈpɪtəmaɪz/ 1A
　　v. 成为……的典范,作为……的缩影
erosion /ɪˈrəʊʒn/ 8A
　　n. 腐蚀
erosional /ɪˈrəʊʒənəl/ 5A
　　adj. 侵蚀的,冲蚀的
erupt /ɪˈrʌpt/ 2A
　　v. 喷发
evacuate /ɪˈvækjueɪt/ 8C
　　v. 撤离,排空
evaporation /ɪˌvæpəˈreɪʃn/ 5A
　　n. 蒸发

evaporative cooling /ɪˈvæpəˌreɪtɪv ˈkuːlɪŋ/ 3B
　　蒸发冷却,蒸发降温
excavation /ˌekskəˈveɪʃn/ 4B
　　n. (对古物的)发掘,挖掘
exhaustibility /ɪɡˌzɔːstəˈbɪləti/ 6A
　　n. 可用尽,耗竭性
exotic /ɪɡˈzɑːtɪk/ 2B
　　adj. 奇异的,异国风情的
expanse /ɪkˈspæns/ 6B
　　n. 宽阔,广阔的区域,膨胀扩张
exploitation /ˌeksplɔɪˈteɪʃn/ 6A
　　n. 开发,开采,(出于私利、不公正的)利用
external /ɪkˈstɜːrnl/ 2B
　　adj. 外部的,外面的
extract /ekˈstrækt/ 2C
　　v. 取出,拔出,提取,提炼

F

far-reaching /ˌfɑːr ˈriːtʃɪŋ/ 3C
　　adj. 影响深远的,波及广泛的
fascinating /ˈfæsɪneɪtɪŋ/ 2A
　　adj. 极有吸引力的,迷人的
fatal /ˈfeɪtl/ 2C
　　adj. 致命的
fault /fɔːlt/ 1C
　　n. 断层
fault strike /fɔːlt straɪk/ 1C
　　断层走向
fauna /ˈfɔːnə/ 6A
　　n. (某地区、环境或时期的)动物群,动物界
federal /ˈfed(ə)rəl/ 8A
　　n. 联邦的,联邦政府的
feldspar /ˈfeldspɑː(r)/ 2A
　　n. 长石
fiber /ˈfaɪbər/ 2B
　　n. 纤维
fine-grained /faɪn ɡreɪnd/ 3A
　　adj. 细粒的

finitude /ˈfɪnɪˌtuːd/ 1A
　　n. 有限,界限,限制
firn /fɪrn/ 5B
　　n. 积雪
fishery /ˈfɪʃəri/ 6A
　　n. 渔业,渔场,水产业
fission /ˈfɪʃn/ 6C
　　n. 裂变,分裂
flashy /ˈflæʃi/ 8A
　　adj. (因巨大等)显眼的
floodwall /ˈflʌdˌwɔːl/ 8A
　　n. 防洪墙,防洪堤
flora /ˈflɔːrə/ 6A
　　n. (某地区、环境或时期的)植物群,植物界
flounder /ˈflaʊndər/ 5C
　　n. 鲆,鲽,比目鱼(同 flatfish)
fluctuation /ˌflʌktjuˈeɪʃən/ 5A
　　n. 波动,变动,起伏现象
fluoride /ˈflɔːraɪdˌˈflʊraɪd/ 5A
　　n. 氟化物

flux /flʌks/ 5A
    n. 流量，流动
forego /fɔːrˈɡoʊ/ 2C
    v. 发生在……之前
fortification /ˌfɔːrtifiˈkeiʃn/ 2C
    n. 碉堡，防御工事
fossil fuels /ˈfɑːsl ˈfjuːəlz/ 3A
    化石燃料
fossil pollen /ˈfɑːsl ˈpɑːlən/ 3A
    花粉化石
fraction /ˈfrækʃn/ 6C
    n. 分数，小数，小部分，微量

fracture /ˈfræktʃər/ 2B
    v. 破裂，折断，瓦解，分裂
fracturing /ˈfræktʃəriŋ/ 7C
    n. 水力压裂，破碎，龟裂
fragmentation /ˌfræɡmenˈteiʃn/ 8C
    n. 破碎，破裂
fuel /ˈfjuːəl/ 3A
    v. 加剧，推动
fungi /ˈfʌndʒai/ 4A
    n. 真菌，菌类，蘑菇（fungus 的复数）
fuse /fjuːz/ 8C
    n. 导火线

G

gamma rays /ˈɡæmə ˈreiz/ 8C
    伽马射线，γ射线
gaseous /ˈɡæsiəs/ 8B
    adj. 气态的，含气体的
gastrointestinal /ˌɡæstroʊinˈtestinl/ 4C
    adj. 胃肠的
geologic /ˌdʒiəˈlɑdʒik/ 3A
    adj. 地质的
geologic isolation /ˌdʒiəˈlɑːdʒik ˌaisəˈleiʃn/ 6C
    地质隔离
geophysical /ˌdʒiːoʊˈfizikl/ 1A
    adj. 地球物理学的
geosphere /ˈdʒioʊsfiə/ 1A
    n. 岩石圈，陆界
geothermal /ˌdʒiːoʊˈθɜːrml/ 6B
    adj. 地热的，地温的
give off /ɡiv ɔːf/ 8B
    放出，发散
glaciation /ˌɡleiʃiˈeiʃn/ 7C
    n. 冰川作用
glaciology /ˌɡleisiˈɑːlədʒi/ 1A
    n. 冰河学，冰川学

glory /ˈɡlɔːri/ 2C
    n. 荣誉
grail /ɡreil/ 6C
    n. 杯，圣杯，大盘，长期以来梦寐以求的东西
grains /ɡreinz/ 5B
    n. 谷粒
granite /ˈɡrænit/ 4A
    n. 花岗岩
granular /ˈɡrænjələr/ 2C
    adj. 颗粒的，粒状的
gravel /ˈɡræv(ə)l/ 2A
    n. 碎石，沙砾
gravity /ˈɡrævəti/ 8C
    n. 重力，地球引力
gravity high /ˈɡrævəti hai/ 1C
    重力高
groundwater /ˈɡraʊndwɔːtər/ 5A
    n. 地下水
gypsum /ˈdʒipsəm/ 2B
    n. 石膏

H

half-life /ˈhæf laif/   6C
   n. 半衰期
halite /ˈhælait/   5C
   n. 岩盐
harness /ˈhɑːrnis/   6A
   v. 控制，利用（以产生能量等）
hauling /ˈhɔːliŋ/   2C
   n. 搬运，拖运
hazard /ˈhæzərd/   8A
   n. 危险，危害
hazardous /ˈhæzərdəs/   6C
   adj. 危险的，有害的
haze /heiz/   3B
   n. 霾
heat island /hiːt ˈailənd/   3B
   n. （城市上空气温偏高的）热岛
helium /ˈhiːliəm/   6C
   n. 氦
helminthiasis /ˌhelminˈθaiəsis/   4C
   n. 蠕虫病，肠虫病
herbaceous /ɜːrˈbeiʃəs/   4B
   adj. 草本的，叶状的
heterogeneity /ˌhetərədʒəˈniːəti/   4C
   n. 异质性，非均匀性
hexagonal /hekˈsægənl/   5B
   adj. 六边的，六角形的
homeostatic /ˌhomiəˈstetik/   1A
   adj. 自我平衡的，原状稳定的
hornblende /ˈhɔːnblend/   2A
   n. 角闪石

human-generated /ˈhjuːmən ˈdʒenəreitid/   3A
   adj. 由人产生的
human-induced /ˈhjuːmən inˈduːst/   3A
   adj. 人为造成的，人类导致的
humus /ˈhjuːməs/   4A
   n. 腐殖质，腐殖土
hydroelectric /ˌhaidrouiˈlektrik/   7A
   adj. 水力发电的
hydrogen /ˈhaidrədʒən/   5C
   n. 氢，氢气
hydrogen dioxide /ˈhaidrədʒən daiˈɑːksaid/   6C
   过氧化氢
hydrologic cycle /ˌhaidrəˈlɒdʒik ˈsaikəl/   5A
   水文循环，水循环
hydrological /ˌhaidrəˈlɑːdʒikəl/   1A
   adj. 水文学的
hydrologist /haiˈdrɒlədʒist/   8A
   n. 水文专家，水文工作者
hydroponics /ˌhaidrəˈpɑːniks/   4B
   n. 水培，无土栽培
hydrosphere /ˈhaidrousfir/   3A
   n. 地球水圈
hydrothermal /ˌhaidrəˈθɜːrml/   7C
   adj. 热液的，热水的
hypothesis /haiˈpɒθəsiz/   3A
   n. 假设（复数为 hypotheses）
hypothesize /haiˈpɒθəsaiz/   1A
   v. 假设，假定

I

i. e. (id est)
　即，也就是

ice caps /ˈais kæp/　5A
　n. 冰盖，冰原，冰冠

ice sheet /ˈais ʃiːt/　5A
　n. 冰原，冰层，冰盖

identification /aiˌdentifiˈkeiʃn/　2A
　n. 辨认，识别

identify /aiˈdentifai/　3C
　v. 查明，确认

igneous /ˈigniəs/　2A
　adj. 火成的

immobilization /iˌmoubələˈzeiʃn/　6C
　n. 使停止流通，固定化，使活动受限

impact /ˈimpækt/　8C
　n. 撞击，冲击力

impactor /ˈimpæktə/　8C
　n. 冲撞器

impending /imˈpendiŋ/　8B
　adj. 即将发生的，迫在眉睫的

imperceptible /ˌimpərˈseptəbl/　3A
　adj. 难以察觉的，察觉不出的

impermeable /imˈpɜːrmiəbl/　7C
　adj. 不可渗透的，透不过的

impurity /imˈpjʊrəti/　2B
　n. 杂质

inadvertently /ˌinədˈvɜːrtəntli/　3A
　adv. 无意地，不经意地

incentive /inˈsentiv/　7B
　n. 激励，刺激

incorporate /inˈkɔːrpəreit/　8C
　v. 包含，吸收，将……包括在内，使并入

indice /ˈindis/　3C
　n. (古法语)指数，标记体

induce /inˈdjʊst/　1B
　adj. 诱发，因某种原因导致

inequity /inˈekwəti/　6A
　n. 不公平，不公正

inert /iˈnɜːrt/　6C
　adj. 惰性的

inert gas /iˌnɜːrt ˈgæs/　3A
　惰性气体

inevitability /inˌevitəˈbiləti/　8C
　n. 必然性，不可避免

inevitable /inˈevitəbl/　8C
　adj. 不可避免的

infancy /ˈinfənsi/　3C
　n. 初期，幼年，婴儿期

infiltration /ˌinfilˈtreiʃn/　4A
　n. 渗透，渗透物

influx /ˈinflʌks/　6B
　n. (人或物的)大量涌入，大量流入，注入

infrared /ˌinfrəˈred/　6B
　adj. 红外线的

infrared radiation /ˌinfrəˈred ˌreidiˈeiʃn/　3B
　红外辐射

inherent /inˈhirənt/　2B
　adj. 内在的，固有的

inhibit /inˈhibit/　3B
　v. 抑制，约束

inhibiting /inˈhibitiŋ/　8A
　adj. 起抑制作用的，抑制的，约束的

initiative /iˈniʃətiv/　1A
　n. 措施，倡议，主动性，积极性

inorganic /ˌinɔːrˈgænik/　6A
　adj. 无机的，无生物的

inrreversible /ˌiriˈvɜːrsəbl/　6B
　adj. 不可逆的，不能取消的，不能翻转的

instantaneously /ˌinstənˈteiniəsli/　8C
　adv. 瞬间地；立即

instrumentation /ˌinstrəmenˈteiʃn/　1A
　n. 使用仪器，仪表化

inter-annual /inˈtɜːr ˈænjuəl/	3C	interrelated /ˌintəriˈleitid/	6B
adj. 年际间的		adj. 相关的,互相联系的	
intercept /ˌintərˈsept/	5A	intersection /ˌintərˈsekʃn/	4C
v. 拦截,阻截		n. 交接(点或线),相交,交汇点(尤指道路)	
interconnectedness /ˌintərkəˈnektidnis/	1A	interval /ˈintərvəl/	8A
n. 互联性		n. 间歇,间隔	
intergrown /ˌintərˈgroun/	2B	intractable /inˈtræktəbl/	6C
adj. 共生的		adj. 棘手的,难处理的	
interlocked /ˌintəˈlɒkt/	5B	intrusive /inˈtruːsiv/	7A
adj. 连锁的		adj. 侵入的,打扰的	
intermittent /ˌintərˈmitənt/	7A	ion /ˈaiən/	2A
adj. 间歇的,断断续续的		n. 离子	
internal /inˈtɜːrnl/	2B	irreversible /ˌiriˈvɜːrsəbl/	1A
adj. 内部的,体内的		adj. 不可挽回的,无法逆转的	
interplay /ˌintərˈplænəteri/	4A	isotope /ˈaisətoup/	1C
n. 相互影响,相互作用		n. 同位素	

K

keyhole /ˈkiːhoul/	8C	kinetic /kiˈnetik/	6B
n. 钥匙孔,锁孔		adj. 运动的,活跃的	
kinematic /ˌkinəˈmætik/	1C		
adj. 运动学的			

L

La Niña /laː ˈniːnjə/	3C	levee /ˈlevi/	8A
拉尼娜现象(在某些年份发生的东热带太平洋变冷)		n. 防洪堤,码头	
		v. 筑堤	
land mass /lænd mæs/	6B	lithospheric /ˌliθəˈsferik/	1C
n. 陆块,地块,大陆块		adj. 石圈的	
latency /ˈleitənsi/	8C	livestock /ˈlaivstɑːk/	6A
n. 延迟,潜伏(期)		n. 牲畜,家畜	
lava /ˈlaːvə/	2A	locus /ˈloukəs/	1A
n. (火山)熔岩,岩浆		n. 地点,所在地	
leftover /ˈleftouvər/	6C	longevity /lɔːnˈdʒevəti/	2C
adj. 剩下的,多余的		n. 持续时间,耐用期限	
legume /ˈliˈgjuːm/	4B	luster /ˈlʌstər/	2B
n. 豆类,豆科植物		n. 光泽,光彩	

M

magma /ˈmæɡmə/ 7C
    n. 岩浆

magma chamber /ˈmæɡmə ˈtʃeɪmbə(r)/ 2A
    岩浆房，岩浆库

magnesium /mæɡˈniːziəm/ 5C
    n. (化学元素)镁

magnetic /mæɡˈnetɪk/ 1B
    adj. 磁的，磁性的，磁化的

magnetic field /mæɡˈnetɪk fiːld/ 1B
    磁场

magnitude /ˈmæɡnɪtuːd/ 1C
    n. 等级，数量

maintenance /ˈmeɪntənəns/ 7A
    n. 维护，保养，维持

manganese /ˈmæŋɡəniːz/ 4B
    n. (化学元素)锰

manifestation /ˌmænɪfeˈsteɪʃn/ 1C
    n. 表现形式

marble /ˈmɑːrbl/ 4A
    n. 大理石

margin /ˈmɑːrdʒɪn/ 8A
    n. 界限，边缘，差额，余地

mass-efficient /mæsɪˈfɪʃnt/ 8C
    adj. 质量效率高的，即轻质量可以获得大能量的

megaton /ˈmeɡətʌn/ 8C
    n. 百万吨级(爆炸能量单位，相当于 100 万 t 黄色炸药的威力)

meltdown /ˈmeltdaʊn/ 6C
    n. 核反应堆堆芯熔毁，(公司、机构或系统的)崩溃

mercury /ˈmɜːrkjəri/ 4B
    n. 汞，水银

metallic /məˈtælɪk/ 6A
    adj. 含金属的，金属制的

metamorphic /ˌmetəˈmɔːfɪk/ 2A
    adj. 变质的

meteorite /ˈmiːtiəraɪt/ 1B
    n. 陨石

meteorologically /ˌmiːtiəraˈdʒɪkli/ 3A
    adv. 从气象学角度看

meteorologist /ˌmiːtiəˈrɑːlədʒɪst/ 3A
    n. 气象学者

meteorology /ˌmiːtiəˈrɑːlədʒi/ 1A
    n. 气象状态，气象学

methane /ˈmeθeɪn/ 3A
    n. 甲烷

mica /ˈmaɪkə/ 2B
    n. 云母

microbial /maɪˈkroʊbiəl/ 4B
    adj. 微生物的，由细菌引起的

microbiome /ˈmaɪkroʊˌbaɪoʊm/ 4C
    n. 微生物组

microorganism /ˌmaɪkroʊˈɔːrɡənɪzəm/ 4A
    n. 微生物，微小动植物

millennia /mɪˈleniə/ 2C
    n. 千年

mineraloid /ˈmɪnərəlɔɪd/ 2A
    n. 准矿石

minute /maɪˈnjuːt/ 2C
    adj. 极小的，微小的，详细的，细致入微的

mitigation /ˌmɪtɪˈɡeɪʃn/ 8C
    n. 减缓，缓解

moderate /ˈmɑːdərət/ 6C; 8C
    v. 缓和，使适中，使(中子)减速
    adj. 中等的，适度的

molecule /ˈmɑːlɪkjuːl/ 3A
    n. 分子

momentum /moʊˈmentəm/ 8C
    n. 动量，动力

monitor /ˈmɒnədər/		8A
v. 监控，监视，监听，检查		
mortar /ˈmɔːrtər/		2C
n. 砂浆，灰浆		
multidimensional /ˌmʌltidiˈmɛnʃənəl/		3A
adj. 多维的		
municipality /mjuːˌnisiˈpæləti/		4C
n. 市政当局，自治市		
muscovite /ˈmʌskəvait/		2A
n. 白云母		
mustard /ˈmʌstərd/		4B
n. 芥末酱，芥末黄		
mycotoxin /ˌmaikoˈtɔksən/		4C
n. 霉菌毒素，真菌毒素		

N

necrosis /neˈkrousis/		4B
n. 坏疽，骨疽		
neutron /ˈnuːtrɑːn/		6C
n. 中子		
nickel /ˈnik(ə)l/		1B
n. （金属）镍		
nitrate /ˈnaitreit/		4B
n. 硝酸盐		
nitrogen /ˈnaitrədʒən/		3A
n. 氮		
nitrous oxide /ˌnaitrəs ˈɑːksaid/		3A
一氧化二氮，笑气		
nominal /ˈnɑːminl/		8C
adj. 名义上的；象征性的		
nonexistent /ˌnɑːnigˈzistənt/		4A
adj. 不存在的		
non-renewable /ˌnɑːn riˈnuːəbl/		6A
adj. 不可再生的，不可更新的		
notoriously /nouˈtɔːriəsli/		7B
adv. 声名狼藉地，臭名昭著地		
notwithstanding /ˌnɑːtwiθˈstændiŋ/		7C
prep. 虽然，尽管		
nuclear fusion /ˈnuːkliər ˈfjuːʒn/		6C
n. 核聚变，核子融合		
nuclei /ˈnukliˌai/		3B
n. 核心，核子		
nucleus /ˈnuːkliəs/		6C
n. 原子核，细胞核，核心，核		
numerical /njuːˈmerikl/		1A
adj. 数字的，用数字表示的		
nutrient /ˈnuːtriənt/		6A
adj. 营养的，滋养的		

O

obsidian /əbˈsidiən/		2A
n. 黑曜石		
occurrence /əˈkʌrəns/		2A
n. 出现，发生		
oceanic ridge basalt /ˌəʊʃiˈænik ridʒ ˈbæsɔːlt/		1C
洋脊玄武岩		
oceanographic /ˌoʊʃənəˈgræfik/		3C
adj. 海洋学的，有关海洋学的		
oceanography /ˌoʊʃəˈnɑːgrəfi/		1A
n. 海洋学		
olivine /ˌɒliˈviːn/		2A
n. 橄榄石，黄绿		
opal /ˈəʊp(ə)l/		2A
n. 欧泊，蛋白石		

词条	位置
optimum /ˈɑːptɪməm/ *adj.* 最优的，最适宜的	4A
orbital /ˈɔːrbɪtl/ *adj.* （行星或空间物体）轨道的	8C
ore /ɔːr/ *n.* 矿石，矿砂	6A
organism /ˈɔːrɡənɪzəm/ *n.* 生物，有机体	5C
orthoclase /ˈɔːθəʊkleɪs/ *n.* 正长石	2A
outcrop /ˈaʊtkrɒp/ *n.* 露头，露出地面的岩层	2A
outlet /ˈaʊtlet/ *n.* 出口，排放管	8A
output /ˈaʊtpʊt/ *n.* 产出	8C
outskirt /ˈaʊtˌskɜːrt/ *n.* 郊区，市郊	3B
outweigh /ˌaʊtˈweɪ/ *vt.* 超过，比……重	7A
oversee /ˌoʊvərˈsiː/ *v.* 监管，监督	6A
overtake /ˌoʊvərˈteɪk/ *v.* 追上，超过	7A
overtop /ˌoʊvəˈtɒp/ *v.* 高出，超出	8A
overwhelm /ˌoʊvərˈwelm/ *v.* 淹没，压倒，击败	8A
overwhelmingly /ˌoʊvərˈwelmɪŋli/ *adv.* 压倒性地，不可抵抗地	6B
oyster /ˈɔɪstər/ *n.* 牡蛎，蚝	5C
ozone /ˈoʊzoʊn/ *n.* 臭氧，臭氧层（ozone layer 的简写）	1A

P

词条	位置
palaeomagnetic /ˌpeɪliəʊmæɡˈnetɪc/ *adj.* 古地磁的	1C
paleoclimate /ˌpeɪlioʊˈklaɪmɪt/ *n.* 古气候，地质气候	3C
paleoclimatology /ˌpeɪliəʊˌklaɪməˈtɒlədʒi/ *n.* 古气候学	3A
paradigm /ˈpærədaɪm/ *n.* 典范，范例，样板，范式	1A
parameter /pəˈræmɪtər/ *n.* 参数，变量	2C
parasitic /ˌpærəˈsɪtɪk/ *adj.* 寄生的	4C
partially /ˈpɑːrʃəli/ *adv.* 部分地，不公平地	8A
pathogen /ˈpæθədʒən/ *n.* 病原体，致病菌	4B
paved area /peɪvd ˈeriə/ 铺筑过的区域	8A
paving/cladding slab /ˈpeɪvɪŋ/ˈklædɪŋ slæb/ 铺路板	2C
pedosphere /ˈpedəsfɪə/ *n.* （地球的）土壤圈	2C
pellet /ˈpelɪt/ *n.* 小球	6C
pennycress /ˈpeniˌkres/ *n.* 菥蓂	4B
periodically /ˌpɪəriˈɒdɪkli/ *adv.* 周期性地；定期地	8A
permafrost /ˈpɜːrməfrɔːst/ *n.* 永久冻土	3A
persistent /pərˈsɪstənt/ *adj.* 坚持不懈的，持续的，反复出现的	6C
pesticide /ˈpestɪsaɪd/ *n.* 杀虫剂，农药	4C
petrological /ˌpetrəˈlɒdʒɪkl/ *adj.* 岩石学的	1C

pharmaceutical /ˌfɑːməˈsuːtɪkl/	2A		n. (化学元素)钾
adj. 制药的,药品的		prairie /ˈpreri/	4A
phosphate /ˈfɑːsfeɪt/	4C	n. (北美的)大草原	
n. 磷酸盐		precedent /ˈpresɪdənt/	8C
phosphorus /ˈfɑːsfərəs/	4B	n. 先例,可援用参考的具体例子,实例	
n. 磷		preceding /prɪˈsiːdɪŋ/	3A
photosynthesis /ˌfoʊtoʊˈsɪnθəsɪs/	6B	*adj.* 先前的,前面的	
n. 光合作用		precipitate /prɪˈsɪpɪteɪt/	2A
photosynthesize /ˌfoʊtoʊˈsɪnθəsaɪz/	5C	v. (使)沉淀	
v. 起光合作用,光能合成,通过光合作用产生		precipitation /prɪˌsɪpɪˈteɪʃn/	1A
photovoltaic /ˌfoʊtəvoʊlˈteɪɪk/	7A	n. 降水(如雨,雪,冰雹)	
adj. 光伏的,光电的		precursor /prɪˈkɜːrsər/	1A
phytoremediation /ˌfaɪtəʊriːˌmiːdiˈeʃən/	4B	n. 先驱,前导	
n. 植物修复		predominate /prɪˈdɑːmɪneɪt/	4A
pigmentation /ˌpɪɡmenˈteɪʃn/	4B	v. (数量上)占优势,占支配地位	
n. 色素淀积,染色		pre-instrumental /ˌpriːˌɪnstrəˈmentl/	3C
pillar /ˈpɪlər/	2C	*adj.* 有仪器检测之前的	
n. 柱子,支柱		prelude /ˈpreljuːd/	8B
pinpoint /ˈpɪnpɔɪnt/	8B	n. 前兆	
v. 确定,精准定位		prescription /prɪˈskrɪpʃn/	6A
plagioclase /ˈpleɪdʒiəˌkleɪz/	2A	n. 处方,药方	
n. 斜长石		pressurize /ˈpreʃəraɪz/	1B
plausible /ˈplɔːzəbl/	1B	v. 增压,施加压力	
adj. 合理的		primordial /praɪˈmɔːrdiəl/	1C
plot /plɑːt/	8A	*adj.* 原始的	
v. 绘制(图表),(在地图上)画出		prior to /ˈpraɪər tə/	2C
plowed land /plaʊd lænd/	6B	*prep.* 在……之前	
耕地		priority /praɪˈɔːrəti/	3C
plume /pluːm/	1C	n. 优先事项,优先	
n. 羽流,上升之物(mantle plume 地幔柱)		probability /ˌprɑːbəˈbɪləti/	3C
polar /ˈpoʊlər/	5A	n. 概率,可能性	
adj. 极地的,来自极地的		project /ˈprɑːdʒekt/	8A
porcelain /ˈpɔːrsəlɪn/	2B	v. 预估,设计,规划	
n. 瓷,瓷器		projection /prəˈdʒekʃn/	3C
pore spaces /pɔːr ˈspeɪsɪz/	5A	n. 估算,预测	
孔隙		propagate /ˈprɑːpəɡeɪt/	1B
porosity /pɔːˈrɑːsəti/	8C	v. 传播	
n. 孔隙度,孔隙率		propagation /ˌprɑːpəˈɡeɪʃn/	3C
potassium /pəˈtæsiəm/	1C	n. 传播,传送,蔓延	

protocol /ˈproʊtəkɔːl/	8C	proxy data /ˈprɑːksi ˈdeɪtə/	3A
n. 议定书，条约草案		代用资料，替代性指标	
protozoa /ˌproʊtəˈzoʊə/	4A	pyrotechnology /ˌpaɪroʊ tekˈnɑːlədʒi/	2C
n. 原生动物		n. 高温技术	
protract /prəˈtrækt/	3C		
v. 延长时间			

Q

quartz /kwɔːts/	2A
n. 石英	

R

radially /ˈreɪdiəli/	1C	relocate /ˌriːˈloʊkeɪt/	8A
adv. 放射状地		v. (使)搬迁，迁移	
radiant /ˈreɪdiənt/	3A	renewability /rɪˌnuːəˈbɪləti/	6A
adj. (热、能量)辐射的		n. 可再生性	
radioactive /ˌreɪdioʊˈæktɪv/	6A	replenish /rɪˈplenɪʃ/	6A
adj. 放射性的，有辐射的		v. 补充，重新装满，补足(原有的量)	
radioactive decay /ˌreɪdioʊˈæktɪv dɪˈkeɪ/	6C	repository /rɪˈpɑːzətɔːri/	6C
放射性衰变		n. 贮藏室，仓库	
radius /ˈreɪdiəs/	1B	reradiation /ˌriːreɪdiˈeɪʃn/	6B
n. 半径		n. 再辐射	
reactor /riˈæktər/	6C	resemble /rɪˈzembl/	2B
n. 反应器，(核)反应堆		v. 像，与……相似	
recurrence /rɪˈkʌrəns/	8A	reservoir /ˈrezərvwɑːr/	5A
n. 重现，复现		n. 水库，蓄水池	
reflection /rɪˈflekʃn/	6B	residence time /ˈrezɪdəns taɪm/	5A
n. (光、热或声音的)反射		停留时间，滞留时间	
refract /rɪˈfrækt/	1B	residual /rɪˈzɪdʒuəl/	4A
v. 使(光线)折射		adj. 残留的，(土壤)残余的	
regolith /ˈregəˌlɪθ/	4A	resilience /rɪˈzɪliəns/	4C
n. 风化层，土被		n. 恢复力，(橡胶等的)弹性	
regression /rɪˈɡreʃn/	3C	resinous /ˈrezɪnəs/	2B
n. (统计)回归		adj. 树脂质的，像树脂的	
reinforce /ˌriːɪnˈfɔːrs/	4A	respiration /ˌrespəˈreɪʃn/	5C
v. 加强，加固		n. 呼吸	

restriction /rɪˈstrɪkʃn/	2B		riprap /ˈrɪpˌræp/	8A
n. (受)限制(状态)，(受)约束(状态)			*n.* (防冲)乱石，(防冲)乱石筑成的地基或堤坝	
retard /rɪˈtɑːrd/	4A		rise /raɪz/	1C
v. 减慢，受到阻滞			*n.* 隆起	
retention /rɪˈtenʃn/	4A		rotate /ˈroʊteɪt/	6B
n. 保持，保留			*v.* 旋转，转动	
retrieve /rɪˈtriːv/	8B		rule-of-thumb /ruːl əv θʌm/	8C
v. 找回，取回，挽回，索回			*n.* 靠经验估计，依据经验的方法	
ridge crest /rɪdʒ krest/	1C		runoff /ˈrʌnˌɔːf/	8A
脊峰线			*n.* 径流量，溢出，泻出	
rift /rɪft/	1C		rural /ˈrʊrəl/	3B
n.; *v.* 断裂，分裂			*adj.* 农村的，乡村的	
rigid (plate) /ˈrɪdʒɪd/	1C			
adj. 刚性的(板块)				

S

sabotage /ˈsæbətɑːʒ/	6C		sediment /ˈsedɪmənt/	3A
v. 蓄意破坏，故意毁坏，捣乱，阻挠			*n.* 沉积物	
saline /ˈseɪliːn/	4B		sedimentary /ˌsedɪˈmentri/	2A
adj. 盐的，含盐的，含镁盐类的			*adj.* 沉积而成的	
sanctuary /ˈsæŋktʃueri/	2C		sedimentation /ˌsedɪmenˈteɪʃn/	5A
n. 宗教圣地，(教堂内的)圣坛			*n.* 沉积(作用)	
saturate /ˈsætʃəreɪt/	4A		seismic /ˈsaɪzmɪk/	1B
v. 浸透，使饱和			*adj.* 地震的，由地震引起的	
n. 饱和脂肪			seismograph /ˈsaɪzməɡræf/	1B
scale up /skeɪl ʌp/	8C		*n.* 地震仪	
增加，扩大，提高			seismometer /saɪzˈmɒmətə/	8B
scratching /ˈskrætʃɪŋ/	2B		*n.* (比 seismograph 精密的)测震仪，地震检波器	
adj. 擦伤，刮痕				
sculpture /ˈskʌlptʃər/	2C		selenium /səˈliːniəm/	4B
n. 雕像，雕塑作品			*n.* 硒	
sea urchin /ˈsiː ɜːrtʃɪn/	5C		self-perpetuating /ˌself pərˈpetʃueɪtɪŋ/	1C
n. 海胆			*adj.* 能使自身永久存在的	
sea ice /siː aɪs/	3A		self-sufficient /ˌself səˈfɪʃnt/	6C
海冰			*adj.* 自给自足的	
seasonality /ˌsiːzəˈnæləti/	3C		sensitivity /ˌsensəˈtɪvəti/	3C
n. 季节性			*n.* 敏感性	

serpentine /ˈsɜːrpəntiːn/	4B	spew /spjuː/	5C
n. 蛇纹石,蛇形之物		*v.* (使)喷出,呕吐	
adj. 蜿蜒的,阴险的		spherical harmonics /ˈsferikl hɑːˈmɒniks/	1C
sewer /ˈsuːər/	8A	球(谐)函数	
n. 排水管,污水管		splinter /ˈsplɪntər/	2B
shatter /ˈʃætər/	1B	*n.* 碎片,微小的东西	
v. (使)破碎,碎裂		splitting /ˈsplɪtɪŋ/	6C
short-wavelength /ʃɔːrt ˈweivleŋθ/	3A	*n.* 分裂	
n. 短波长		spreading rate /ˈspredɪŋ reit/	1C
silica /ˈsilikə/	5C	扩张速率	
n. 二氧化硅,硅土		spurge /spɜːrdʒ/	4B
silicate /ˈsilikeit/	2A	*n.* 大戟,大戟树	
n. 硅酸盐		stakeholder /ˈsteikhoʊldər/	4C
silicon /ˈsilikən/	1B	*n.* 股东,利益相关者	
n. (化学元素)硅		stall /stɔːl/	6C
simulation /ˌsimjuˈleiʃn/	3C	*v.* 暂缓,搁置,停顿	
n. 模拟		stationarity /ˌsteiʃənˈærəti/	3C
skew /skjuː/	8B	*n.* 平稳性,稳定性	
v. 偏离,歪斜		stratospheric /ˌstrætəˈsfirik/	1A
slope /sloʊp/	8A	*adj.* 平流层的,同温层的	
n. 斜坡,坡度,坡地		streak /striːk/	2B
sodium /ˈsoʊdiəm/	2A	*n.* 条纹,条痕	
n. 钠		subsequent /ˈsʌbsikwənt/	2C
solar radiation /ˈsoʊlər ˌreidiˈeiʃn/	3A	*adj.* 随后的,接着的	
n. 太阳辐射		subsidize /ˈsʌbsiˌdaiz/	8A
soluble /ˈsɑːljəbl/	4A	*v.* 资助,补助,给……发津贴	
adj. 可溶的,可解决的		substance /ˈsʌbstəns/	6A
solution /səˈluːʃ(ə)n/	2A	*n.* 物质,材料	
n. 溶解过程		substantially /səbˈstænʃəli/	3A
sophisticated /səˈfistikeitid/	2B	*adv.* 很大程度地	
adj. 复杂巧妙的,先进的,精密的		subsurface /ˈsʌbˌsɜːrfis/	7C
spatial variability /ˈspeiʃl ˌveriəˈbiləti/	3C	*adj.* 地下的,表面下的	
n. 空间变异		subtract /səbˈtrækt/	6B
spearhead /ˈspirhed/	7C	*v.* 减去,删减,扣除	
v. 带头,做先锋		sulfate /ˈsʌlfeit/	5C
specification /ˌspesifiˈkeiʃn/	8C	*n.* 硫酸盐	
n. 规范,规格		sulfur /ˈsʌlfər/	1B
spectrum /ˈspektrəm/	8C	*n.* 硫磺	
n. 范围			

sulfur dioxide /ˌsʌlfər daɪˈɑːkˌsaɪd/	8B		sustainable /səˈsteɪnəb(ə)l/	6A
二氧化硫			adj.(自然资源)可持续的,不破坏环境的	
surplus /ˈsɜːrplʌs/	5B		sustainable development	6A
adj.过剩的,剩余的,多余的			/səˈsteɪnəb(ə)l dɪˈveləpmənt/	
n.过剩,剩余,过剩量			可持续发展	
survey /ˈsɜːrveɪ/	6A		synoptic /sɪˈnɑptɪk/	3C
v.测量,勘测			adj.天气的	
sustainability /səˌsteɪnəˈbɪləti/	1A			
n.持续性,能维持性				

T

tailored /ˈteɪlərd/	8C		topography /təˈpɑːgrəfi/	1C
adj.专门设定的			n.地形学,地貌学	
talc /tælk/	2B		torrential /təˈrenʃl/	5C
n.滑石			adj.(水)奔流的,(雨)倾泻的,如注的,猛烈的	
teleconnection /ˈtelɪkəˌnekʃ(ə)n/	3C			
n.遥相关(用于确定年代)			toxic /ˈtɑːksɪk/	6C
temporal /ˈtempərəl/	4C		adj.有毒的,引起中毒的	
adj.时间的,与时间有关的			toxicity /tɑːkˈsɪsəti/	7C
terawatt /ˈterəwɑːt/	6B		n.毒性	
n.(电功率单位)太(拉)瓦,万亿瓦			toxin /ˈtɑːksɪn/	4C
terrain /təˈreɪn/	3B		n.毒素,毒质	
n.地形,地势			trace gases /treɪs ˈgæsɪz/	3A
terrestrial /təˈrestriəl/	1A		微量气体	
adj.地球的,地球上的			tractor /ˈtræktər/	8C
thermal /ˈθɜːrml/	3B		n.牵引器	
adj.热的,由热引起的,由温度变化引起的			trajectory /trəˈdʒektəri/	8C
tholeiitic /ˌθəʊliːˈɪtɪk/	1C		n.轨迹;轨道	
adj.拉斑玄武岩的			transdisciplinary /ˌtrænsˈdɪsəplənəri/	1A
tillage /ˈtɪlɪdʒ/	4C		adj.跨学科的,学科间的	
n.耕作,耕种			transfer /trænsˈfɜːr/	6B
tiltmeter /ˈtɪltˌmiːtə/	8B		v.转移,搬迁,转移	
n.测斜器,地面倾斜度测量仪			transformation /ˌtrænsfərˈmeɪʃn/	6C
toluene /ˈtɑːljuˌiːn/	4C		n.(彻底或重大的)改观,变化,转变,核的转换	
n.甲苯			transition /trænˈzɪʃn/	7B
tonnage /ˈtʌnɪdʒ/	7A		n.过渡,转变	
n.吨位,总质量,总吨数			transparent /trænsˈpærənt/	3A
			adj.可穿透的,(热、电磁波)可通过的	

transpiration /ˌtrænspɪˈreɪʃn/ 5A
    n. 蒸发，散发，蒸腾作用

tremendous /trəˈmendəs/ 6C
    adj. 巨大的，极大的，令人望而生畏的，可怕的

tremor /ˈtremər/ 8B
    n. 轻微地震；小震

trench /trentʃ/ 1C
    n. 海沟

tributary /ˈtrɪbjəteri/ 8A
    n. （流入大河或湖泊的）支流

triple junction /ˈtrɪp(ə)l ˈdʒʌŋkʃn/ 1C
    三向连接构造，三联点

troposphere /ˈtrɒpəsfɪə(r)/ 3A
    n. 对流层

truncation /ˈtrʌnkeɪʃn/ 3C
    n. 截断

tundra /ˈtʌndrə/ 4A
    n. 苔原，土地带

turbine /ˈtɜːrbaɪn/ 6C
    n. 涡轮机，汽轮机

turbulent /ˈtɜːrbjələnt/ 7A
    adj. 骚乱的，湍流的

U

undermine /ˌʌndərˈmaɪn/ 7A
    v. 逐渐削弱，故意破坏……的形象，在……下面挖，从根基处损坏

underpin /ˌʌndərˈpɪn/ 1A
    v. 支持，巩固，构成……的基础

undulating /ˈʌndʒəleɪtɪŋ/ 4A
    adj. 波状的，波浪起伏的

unglazed /ʌnˈɡleɪzd/ 2B
    adj. 未上釉的

unreasonably /ʌnˈriːznəbli/ 8A
    adv. 不合理地

unregulated /ʌnˈreɡjuleɪtɪd/ 6A
    adj. 未受控制的，无管理的，未经调节的

unreliable /ˌʌnrɪˈlaɪəbl/ 2B
    adj. 不可靠的，靠不住的

unrest /ʌnˈrest/ 6A
    n. 不安，动荡的局面，不安的状态

uplift /ˈʌplɪft/ 6B
    n. （地壳的）隆起，举起，抬起

upstream /ˌʌpˈstriːm/ 8A
    adj. 上游的

uranium /juˈreɪniəm/ 6A
    n. （放射性化学元素）铀

urban /ˈɜːrbən/ 3B
    adj. 城市的，城镇的

utilize /ˈjuːtəlaɪz/ 6A
    v. 利用，使用

V

vanish /ˈvænɪʃ/ 8B
    v. 消失

variability /ˌveriəˈbɪləti/ 3A
    n. 可变性，变化性

variable /ˈveriəbl/ 3A
    adj. 易变的，多变的

variance /ˈveriəns/ 3C
    n. 方差，变化幅度，差额

variation /ˌveriˈeiʃn/ 3A
    n. 变化，变动
variety /vəˈraiəti/ 3A
    n. 多样化，变化
vary /ˈveri/ 3A
    v. 变化
vegetation /vedʒəˈteiʃn/ 4A
    n. (总称)植物，植被
velocity /vəˈlɑːsəti/ 1B
    n. 速度
vent /vent/ 1C
    n. 通风口，火山口，(空气、气体、液体的)出口
versatility /ˌvɜːrsəˈtiləti/ 4C
    n. 多功能性，多才多艺
vibrate /vaiˈbreit/ 2A
    v. (使)震动，(使)颤动
viscous equation /ˈviskəs iˈkweiʒ(ə)n/ 1C
    黏性流方程
vital /ˈvaitl/ 2C
    adj. 至关重要的，必不可少的
vitreous /ˈvitriəs/ 2B
    adj. 玻璃状的，透明的
void /vɔid/ 1C
    n. 孔隙
volcanic /vɑːlˈkænik/ 3A
    adj. 火山的，由火山引发的
volcanic eruption /vɑːlˈkænik iˈrʌpʃ(ə)n/ 3A
    火山喷发
vulnerability /ˌvʌlnərəˈbiləti/ 3C
    n. 脆弱性

W

water scarcity /ˈwɔːtər ˈskersəti/ 3C
    水短缺，水荒
water vapor /ˈwɔːtər ˈveipər/ 3A
    水蒸气
waterlog /ˈwɔtəlɔg/ 4A
    v. 使(船等)进水，使浸透水
wedge /wedʒ/ 7C
    n. 楔形物，三角形物
weir /wir/ 8A
    n. 堰，拦河坝，导流坝
withstand /wiðˈstænd/ 2C
    v. 经受住，承受住

Y

yield /jiːld/ 8C
    n. 产出，收益